GEOLOGY
OF NOVA SCOTIA

Martha Hickman Hild and Sandra M. Barr

Other field guides by Boulder Publications

Geology of Newfoundland by Martha Hickman Hild

Birds of Newfoundland by Ian Warkentin and Sandy Newton

Edible Plants of Atlantic Canada by Peter J. Scott

Edible Plants of Newfoundland and Labrador by Peter J. Scott

Trees & Shrubs of the Maritimes by Todd Boland

Trees & Shrubs of Newfoundland and Labrador by Todd Boland

Whales and Dolphins: Atlantic Canada and Northeast United States by Tara S. Stevens

Wildflowers of Newfoundland by Peter J. Scott

Wildflowers of Nova Scotia by Todd Boland

Gardening guides by Boulder Publications

Atlantic Gardening by Peter J. Scott

Newfoundland Gardening by Peter J. Scott

GEOLOGY
OF NOVA SCOTIA

Martha Hickman Hild and Sandra M. Barr

Library and Archives Canada Cataloguing in Publication

Hild, Martha Hickman, author
    Geology of Nova Scotia: field guide / Martha Hickman Hild, Sandra M. Barr.

Includes bibliographical references and index.
ISBN 978-1-927099-43-8 (pbk.)

1. Geology--Nova Scotia--Guidebooks. I. Barr, Sandra M., 1946-, author II. Title.

QE190.H54 2015   557.16   C2014-901637-9

Published by Boulder Publications
Portugal Cove-St. Philip's, Newfoundland and Labrador
www.boulderpublications.ca

© 2015 Martha Hickman Hild, Sandra M. Barr

Book concept: Martha Hickman Hild
Editor: Stephanie Porter
Copy editor: Iona Bulgin
Cover design and page layout: Sarah Hansen & Jessie Meyer

Front cover: Peggys Cove
Back cover: Fourchu Head

Printed in China

Newfoundland Labrador

We acknowledge the financial support of the Government
of Newfoundland and Labrador through the Department
of Tourism, Culture and Recreation.

Canada

We acknowledge financial support for our publishing
program by the Government of Canada and the Department
of Canadian Heritage through the Canada Book Fund.

Dedication

To the people of Nova Scotia, stewards of a unique geological heritage.

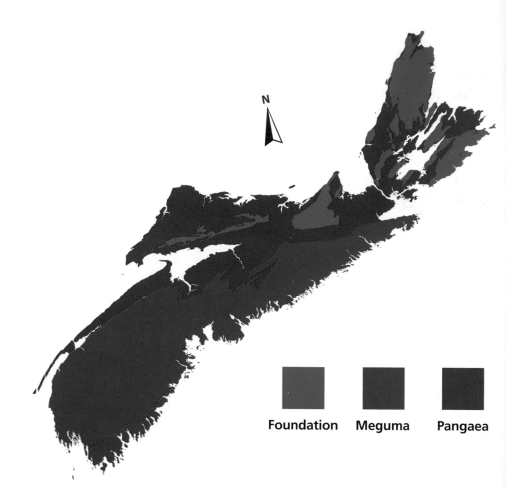

Foundation   Meguma   Pangaea

Table of Contents

Authors' Preface

In a famous story about the beginnings of modern geology, in 1788 James Hutton, James Hall, and John Playfair—three well-known founders of the science—all pile into a little sailing skiff and set off to spend the day on a beach exploring rock outcrops.

One reason the story is so widely known is the effect of this outing on Playfair, who reported that in his excitement he felt positively "giddy" after seeing the rocks with his own eyes and contemplating the story they told. In short, he had a wonderful time.

That day on the beach set an important precedent. Still today many geologists feel that a field trip is the most effective, and certainly the most enjoyable, way to share their insights. It's a popular activity at the heart of our science.

We wrote this book because we think there is absolutely no reason why all that fun should be the sole domain of professionals. If you like to tromp around out of doors and like it even more when you have a purpose in mind, then this book is for you. *Geology of Nova Scotia* will help you find, understand, and enjoy the rocks at a variety of sites.

The book describes 48 sites of primary interest as well as additional, related outcrops and museums—more than 75 locations in all. In our search for suitable sites we have travelled Nova Scotia from end to end, looking for locations that are scenic, accessible, and representative of the province's many geographic regions, as well as its geological past. We have visited each of the sites mentioned in this book as well as many others, selecting only those that we found pleasant and engaging.

Finding and understanding rock outcrops of interest can be a challenge. Using other geology-themed maps or lists, you may wonder: Will the site suit my interests? How long will it take to get to the outcrop? Where exactly is this outcrop, anyway? Once I find it, how will I know what the rocks are telling me?

Our solution was to adopt a relatively new format, one developed by Martha and used for the award-winning field guide *Geology of Newfoundland*. It uses a series of structured, four-page site descriptions (six pages for sites with multiple points of interest) to provide all the information you need to have a good time looking at rocks. In keeping with other guides from Boulder Publications, the size and robustness of the book suit its intended purpose as your companion in the field.

Traditionally, geological field trips are organized around a story told by the rocks about the Earth, its processes and history. To mirror that tradition, we've organized this book into a series of three interwoven narratives about Nova Scotia's geologic past.

Much like the history of its founder populations of Mi'kmaq, Scottish, Acadian, and African descent, the province's geological history has been defined by the joining of diverse, widely distributed segments of the Earth's crust into a common whole. That geological heritage continues to shape the province today, from its landforms to its economy.

If you use this book to visit outcrops in the field, we hope you'll find, as we do, that tracking down Nova Scotia's geological past can provide wonderful recreation. The rocks tell a fascinating story and are well exposed in the province's beautiful parks, shorelines, and other public spaces. You can begin your adventures today.

Acknowledgements

Many people assisted us in bringing this book to life. Alan Macdonald and Rob Raeside provided advice, encouragement, and geological insights throughout the preparation of the book. They also assisted us by reviewing a draft of the book and visiting a few of the sites for us to answer questions about their features. Their helpful feedback had a significant impact on the quality of our text.

Chris White suggested several of the sites, helped clarify field relations, and served as a sounding board for our discussions. Dave Hild provided assistance in the field and spent many hours carefully quality checking specific details. Their contributions were invaluable and we offer them a sincere thank you.

Several individuals contributed images; for their generosity we thank Rob Fensome, Sophie LeLeu, Alan Macdonald, Jonathan Shute, and Chris White.

Knowledge about the geology of Nova Scotia is continually evolving and has been the work of many people. Members of the geological surveys of Canada and of Nova Scotia, faculty and students of the region's universities, and others who have mapped and studied the province's rock formations have all contributed to the information presented here. In particular, we thank Sandra's students over the past nearly 40 years who have shared her enthusiasm and helped her to better know and understand the rocks.

Boulder Publications' capable staff guided the publication process, carefully rendering our content onto the page and arranging for the book's high-quality production. We particularly thank Stephanie Porter, Jessie Meyer, Sarah Hansen, and Iona Bulgin for their work.

Grants from the Canadian Geological Foundation and the Atlantic Geoscience Society supported the field collaborations that were so essential to the success of our project. We are indebted to both these organizations for their contributions.

How to Use This Guide

What's in the Guide

This book is organized as a journey through time and aims to include all the "provisions" you'll need to make the trip. Here's a brief summary.

Sample Pages

Pages 4 and 5 provide a graphical key to the format of the field guide's 48 site descriptions.

Geology Basics

These pages are for readers who want basic background information about geology. How do geologists measure and describe geologic time? What are the main kinds of rocks and how do they form? How do tectonic plates interact? What forces shaped Nova Scotia's geological history? These topics are briefly reviewed at the front of the book.

Resources

For the extra curious, following Geology Basics is a list of geology-themed museums and interpretive centres as well as print and online resources you can explore to learn more.

Trip Planner

On pages 24 and 25 is a Trip Planner you can use to select sites and plan your itinerary based on what interests you, what parts of the province you'll be visiting, how far you want to hike, and other preferences. The Trip Planner lists the 48 sites covered in the book as well as some geology-related museums and interpretive centres.

Section at a Glance

For this book, Nova Scotia's geological history is presented in three sections—Foundation, Meguma, and Pangaea. Each section begins with a summary that includes a map of the sites, a list reviewing what you can experience at each site, and a brief account of geological events.

Site Descriptions

The core of the book describes 48 sites of geological interest. The sequence begins with Nova Scotia's oldest rocks and ends with its youngest. For each site, photographs, maps, diagrams, and text provide information about why the site is interesting, how to get there, what to look for, and what the outcrop means for the geological history of the province.

Summary Timelines

A two-page graphical summary plots all the sites on a geological timescale.

Glossary

Words that may be unfamiliar are explained at the back of the book.

Index

An alphabetical index of place names includes all locations of geological interest mentioned in the text. It contains more than 100 entries in all.

Legends

Book Sections

Throughout the book, specific colours are associated with each of the three sections describing Nova Scotia's geological past.

Foundation   Meguma   Pangaea

Map Symbols

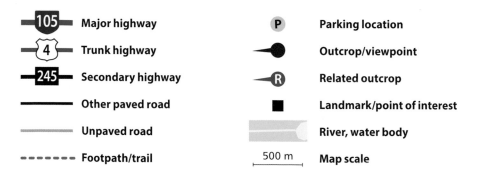

105	Major highway	**P**	Parking location
4	Trunk highway	●	Outcrop/viewpoint
245	Secondary highway	**R**	Related outcrop
	Other paved road	■	Landmark/point of interest
	Unpaved road		River, water body
- - - - -	Footpath/trail	500 m	Map scale

NOTE: All road maps are oriented with North at the top. All geologic maps are oriented with North rotated 10° to the left (westward), unless otherwise indicated.

Icons

NOTE: Icons relevant to the site are printed in a dark, solid colour. Icons not applicable are printed in a lighter tone.

 Walking distance – This is the approximate total walking distance for the site, that is, the distance from the parking location to the outcrop and back.

 Water level – This outcrop is best viewed in low water conditions (low tide or low lake and river levels). Check tide tables or other sources when planning your visit. NOTE: This icon refers to normal tides and normal seasonal water-level fluctuations only. Storms or other unusual conditions affecting water levels may pose safety and access issues even for sites not marked with this icon.

 Seasonal – Access to this outcrop is seasonal due to park closure. NOTE: This icon is not intended to indicate the effect of weather, road conditions, or other seasonal factors on site access.

 Park – This site is located within a national, provincial, or community park, wilderness, or other protected area. It is your responsibility to be aware of any park rules and policies affecting your visit.

 Cost – There is a fee to visit this site and/or the park in which it is located.

Also Note

- For each site, a map number and name are listed under the heading 1:50,000 Map. These identify the National Topographic System of Canada (NTS) map on which the site appears.

- All latitude-longitude readings in the field guide are given in the same format, for example, N45.90616 W59.95734. The letter prefix refers to the hemisphere (northern latitude and western longitude) and should be included when entering the coordinates into a GPS device or internet mapping utility such as Bing™ Maps, Google Maps™, or Google Earth™.

- In many of the photographs, a black and white metric scale is visible. The scale is 10 centimetres long; its smallest subdivisions are 1 centimetre square.

Sample Pages

Site location in easy-to-reference heading, colour-coded by zone.

Local scenery to help you recognize the site when you arrive.

Non-technical **description** of significance and features at the site.

Nearly horizontal layers of orange-red Jurassic sandstone brighten the shore at Five Islands Provincial Park.

Starting Over Again
Early Jurassic River and Lake Deposits

Jurassic. More than any other geologic time period, it evokes specific, dramatic images, thanks in part to the entertainment industry. Dense foliage munched by thundering herbivores—darting, wily velociraptors ... Not far off the mark for late Jurassic times, about 150 million years ago, such scenes earn the period its nickname "age of the dinosaurs."

Hang on, though. Pause to adjust your imagination. The rocks at this site formed quite early in the Jurassic period, about 190 million years ago. Not long before, at the Triassic-Jurassic boundary, nearly half the Earth's known species had become extinct and the region had been smothered in thick sheets of hot lava. The landscape was in recovery mode. Nobody's footsteps thundered, not yet.

The sedimentary features and fossils of these rocks, Nova Scotia's youngest, tell of a changing world. The climate was becoming milder and wetter, but the landscape was still dry much of the time. Plants and animals were repopulating Pangaea, evolving toward the well-known outcome. Something of a cliffhanger, the province's brief but informative Jurassic chapter is aptly nicknamed "dawn of the dinosaurs."

242

1

On the Outcrop

Different weathering rates highlight alternating layers of sandstone and mudstone in the cliff. Near the base, note the thin, lens-shaped area of lake or pond deposits, right of centre (see detail).

Outcrop Location: N45.39309 W64.06159

The thick layers in the cliff were deposited by braided streams—a complex, criss-crossing system of channels that quickly deposited large volumes of sandstone. During floods, widespread layers of mudstone formed as finer sediment settled out slowly.

An eye-catching feature here is a lens-shaped area several metres long near the base of the cliff. The slightly recessed rock is striped with red and grey layers a few centimetres thick. They were deposited in an area of standing water, for example, on the margin of a lake or in a pond that was refilled with flood water from time to time.

About 20 metres south of this feature is a sandstone layer containing tablet-shaped fragments of mudstone. Known as rip-up clasts, the fragments formed as fast-flowing, sand-laden water rushed over the surface of an older mud layer, breaking it into pieces.

Lake or pond deposits.

Image of the outcrop to help you find and interpret the rocks.

GPS waypoint for the outcrop location —nature's geocache. Provided in decimal degrees so you can easily enter it into your GPS device or web browser.

Description of the outcrop emphasizing what to look for.

244

1000	900	800	700	600
		J_1	J_2	J_3

3

Getting There

Driving Directions

Along Highway 2 about 9 kilometres west of Economy, watch for signs to Five Islands Provincial Park and turn (N45.40771 W64.02163) south onto Bentley Branch Road. Follow the road for about 3 kilometres to the park office. Continue down and then up a hill. Near the crest of the hill, fork right onto a gravel road for the day-use picnic area and parking location.

Where to Park

Parking Location: N45.39556 W64.06091

Park in the gravel parking lot for the day-use picnic area.

Walking Directions

From the parking location, walk north down a grassy slope. Where it meets the shore, as conditions allow, cross onto the beach and turn left (south). Walk south following the shoreline for about 375 metres to the first outcrop. Continue 350 metres farther for the second outcrop. For a dramatic view farther along the shore, continue 250 metres onto the rocky saddle at the Old Wife. Optionally—as conditions allow—climb across or walk around the saddle to continue southeast along the beach.

Notes

This site lies within the boundaries of Five Islands Provincial Park. The shore is subject to high Fundy tides. Low tide conditions are required both for access and for a safe return. Staff at the park office can advise how best to plan your shoreline walk.

1:50,000 Map	Provincial Scenic Route
Parrsboro 021H08	Glooscap Trail

2

Thumbnail map showing site location.

Detailed written **directions** for driving, parking (including GPS waypoint), and walking.

Road map showing the **route** to the outcrop from the nearest highway (for Legend, see page 2).

At-a-glance **information** to help you plan your trip: key site characteristics plus topographic map and tourism route details.

By the Way

FYI

- Nova Scotia's Jurassic rocks contain the richest collection of early dinosaur bones in North America and the oldest known dinosaur bones in Canada (see Exploring Further).
- The Jurassic fossils of Nova Scotia include pollen and spores indicating a variety of vegetation supporting the food chain. Though the Atlantic Ocean had begun to open, land animals could still roam Pangaea. Jurassic fossils found in Nova Scotia have close relatives in the western United States, China, and South Africa.

The shore at Wasson Bluff with Five Islands in the distance.

Related Outcrops

Exposures of Jurassic rock in Nova Scotia are limited to a few small areas along the eastern Bay of Fundy. Just east of Five Islands is Wasson Bluff, a site where thousands of Jurassic fossil tracks and bones have been discovered, including those of ancient crocodiles, lizards, and small dinosaurs.

To visit Wasson Bluff, from Parrsboro follow Two Islands Road about 9 kilometres eastward to a gravel parking area (N45.39555 W64.22855) on the south side of the road. The trail head is marked by a sign, "Dawn of the Dinosaurs," including a map of the shore. The site is protected by Nova Scotia's Special Places Act and no collecting is allowed.

Exploring Further

Fundy Geological Museum, 162 Two Islands Road, Parrsboro (N45.39961 W64.32387), fundygeological.novascotia.ca. The museum offers displays and activities related to the volcanic minerals and fossil discoveries of the Parrsboro area, including its remarkable yield of early Jurassic dinosaur fossils.

500   400   300   200   100

C O S D C P T J K

4

Two **extra pages** for some sites highlighting additional features of interest.

Research findings that help you appreciate the outcrop's significance.

Information about **other sites** with similar rocks (may include a map, additional image, or GPS waypoints).

Resources for exploring topics related to the site.

Timeline showing the outcrop's place in geologic time (also see pages 8 and 9).

Exploring the Sites

Safety

You can't experience geology directly without going outdoors. All the sites in this book have been visited safely by many people. They are places where residents, tourists, geologists, students, and others go to enjoy the beauty of the province and examine interesting rocks. However, weather, tides, and other conditions can render any site hazardous temporarily. Only you can decide whether it is safe to visit a specific site on a specific day and whether your state of preparedness is appropriate for the conditions.

Navigating

Each site description provides information to help you find the outcrop easily. Road maps and written directions indicate one possible route from the nearest highway to the parking location; further instructions describe how to reach the outcrop. Additional details help you find the site on topographic or tourism maps.

All field readings of GPS latitude and longitude reference World Geodetic System datum WGS84. If you enter them into your own GPS device, web browser, or other application, keep in mind that results may vary depending on your GPS device settings, map projections, application preferences, and other factors. The readings are not intended, and should never be used, as a substitute for attentive real-world navigation.

The parking locations given are example locations where parking was possible at the time of publication. Please exercise common sense and courtesy at all times when parking your vehicle, based on the conditions you encounter. The listed outcrop location marks the position of the rock described in the text or a vantage point from which significant features can be viewed.

Written directions to the sites include words and phrases such as "northward," "to the east," or "southwest." They are used only in a general sense to clarify correct choices at forks, intersections, and other turning points. They are not intended as precise orienteering instructions. Driving and hiking distances are approximate and are provided only as an aid in planning your trip and in finding the sites and outcrops.

Preserving Outcrops

The geology of Nova Scotia is unique. If you use this field guide to visit sites, please apply the principles of Leave No Trace Canada (www.leavenotrace.ca) to preserve Nova Scotia's natural beauty and geological treasures for others to enjoy. Don't hammer outcrops or remove material—photographs are the best way to capture your experiences.

Protected Sites

Certain areas of Nova Scotia are protected by law from any hammering or rock collecting. In addition, it is against the law to remove fossil material from any outcrops in Nova Scotia without a permit. The second page of each unit (Getting There) identifies sites protected at the time of publication. However, protected status of sites is subject to change. The Nova Scotia Department of Natural Resources is the best source for current information on the province's Restricted and Limited Use Lands (www.novascotia.ca/parksandprotectedareas/plan/interactive-map).

Geology Basics

Geologic Time

Superposition and Cross-Cutting

The features of an outcrop can reveal the relative ages of geologic events, that is, the order in which they occurred.
The principles of superposition and cross-cutting relations allow you to reconstruct a sequence of events even though they happened long ago.

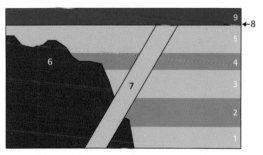

A simplified rock outcrop showing the order in which features formed: 1–5, sedimentary layers; 6, igneous intrusion; 7, cross-cutting dyke; 8, erosion surface; 9, a younger sedimentary layer.

Superposition. Sedimentary rocks are deposited in horizontal layers, with younger layers on top of older layers.

Cross-cutting. Any feature (intrusion, fault, erosion surface) that cuts across or truncates other features is younger than all the features it cuts across.

Fossil Species and Radiometric Data

Geologists use two types of evidence—fossil species and radiometric data—to tell when rocks formed and geologic events happened.

Fossil species. As life evolved in the geologic past, new species appeared, existed for a time, and became extinct. Based on close study of the appearance, distribution, and extinction of fossil species preserved in rock, the international geological community has agreed on standards that define which fossils belong to which geologic period. This information allows a rock layer to be assigned an age based on the fossils it contains.

Radiometric data. Some rocks and minerals contain small amounts of radioactive atoms. Radioactive decay converts an unstable atom (called the parent isotope) into a stable atom (called the daughter isotope). For example, radioactive atoms of uranium decay to form stable atoms of lead. By measuring the amount of parent and daughter isotopes very precisely, geologists can calculate the age of the rock based on the known rate of radioactive decay.

Proterozoic Eon

Late Proterozoic Era

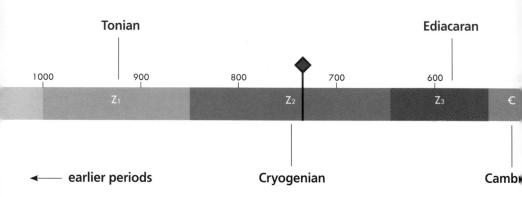

Tonian

Ediacaran

1000   900   800   700   600

Z₁   Z₂   Z₃   €

←— earlier periods   Cryogenian   Camb

The Book's Timeline

On the third and fourth pages of each site description is a ribbon of colour like the one shown above. The numbers on the timeline mark off intervals of 100 million years before the present day; the colours mark the boundaries of geologic periods and eras; and the letters are abbreviations for the names of the periods (except Cenozoic, which is an era, not a period).

Eras are shown on the timeline by colours: shades of brown for the Neoproterozoic (or late Proterozoic) era; shades of blue for the Paleozoic era; shades of green for the Mesozoic era; and yellow for the Cenozoic era.

Each site description includes a marker on the timeline showing the age of the rock or event of interest at that site. Events described in this book span a period of nearly 900 million years, from 1,080 to 190 million years ago. In the example above, the marker is placed at 734 million years ago.

Exploring Further

Canadian Federation of Earth Sciences. *Four Billion Years and Counting*, Ch. 3, "It's About Time" (pp. 38–44).

Edwards, Lucy and John Pojeta, Jr. *Fossils, Rocks, and Time* (Online Edition). U.S. Geological Survey website, pubs.usgs.gov/gip/fossils.

Phanerozoic Eon

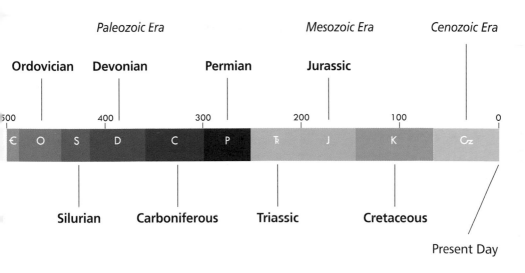

Four Eons of Geologic Time

The book's timeline shows only part of the history of the Earth—our planet formed about 4,550 million years ago. Geologists divide geologic time into eons, which are subdivided into eras and periods. Here's a quick summary of what happened during each eon of geologic time:

Hadean eon (4,550 to 4,000 million years ago). Very few rocks or minerals survive from the Hadean eon. The Earth's crust was hot, mobile, and bombarded by meteorites.

Archean eon (4,000 to 2,500 million years ago). Tectonic plates and early continents formed; primitive single-celled micro-organisms appeared in the sea.

Proterozoic eon (2,500 to 541 million years ago). Areas of stable continental crust grew and mountain belts formed during a series of continental collisions. The supercontinent Rodinia formed, then broke up. Oxygen levels increased in the air and sea, allowing early complex life forms to evolve.

Phanerozoic eon (541 million years ago to the present). Continents re-assembled to form the supercontinent Pangaea. Pangaea broke up as present-day continents moved to their current positions. Animals and plants evolved to inhabit both sea and land.

Rock Types

Identifying Rocks

"What kind is it?"

That's the first question that comes to mind when you see a rock in the field. Correctly identifying the type of rock you are looking at is the first step in reading the rock's story about the geological past.

Nova Scotia provides easy access to a remarkable variety of rock types, formed in a wide range of conditions—it's a natural showcase of Earth materials.

There are plenty of general guides to rock identification in print and online. This brief review uses examples from the sites described in this book.

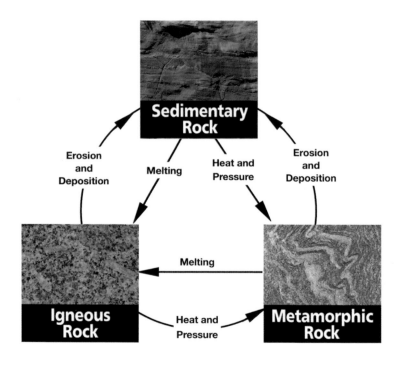

The rock cycle shows how igneous, sedimentary, and metamorphic rocks can be transformed and recycled into other rock types when exposed to new conditions. Changing conditions can be caused by uplift, burial, the movement of fluids in the crust, and other events as the Earth's tectonic plates interact.

Igneous Rocks

Igneous rocks form when molten rock (magma) rises through the crust, cools, and crystallizes. Igneous rocks are classified based on the minerals they contain, then subdivided into two varieties: intrusive (crystallized slowly underground and thus coarse-grained) and extrusive (erupted onto the surface, crystallizing quickly and thus fine-grained).

Felsic igneous rocks include granite (intrusive) and rhyolite (extrusive). They contain mostly light-coloured minerals (quartz and feldspar), often with small amounts of dark minerals such as biotite or amphibole (sites 3, 4, 7, 12, 15, 16, 24, 29, 30). True granite contains potassium feldspar, plagioclase, and quartz. However, the term "granitic" is used informally to refer to a wide variety of igneous rocks dominated by any combination of these light-coloured minerals.

Granite, Green Cove (site 15).

The most common intrusive intermediate rocks (site 11) are granodiorite and diorite (intrusive) or their fine-grained equivalents, dacite and andesite (extrusive). They contain a mixture of light and dark minerals.

Mafic igneous rocks contain approximately equal amounts of plagioclase and dark minerals, typically pyroxene. The plagioclase is calcium-rich and typically grey (sites 4, 6, 16, 24, 36, 45, 46, 47). Gabbro (intrusive) and basalt (extrusive) both have a similar mixture of minerals but differ in the size of the mineral grains— easily visible in gabbro to microscopic in basalt.

Some igneous rocks do not fit this simple classification. Anorthosite, for example, is made almost exclusively of calcium-rich plagioclase (site 1). Most anorthosite intrusions are Proterozoic in age, and geologists agree that they formed very deep in the crust. However, their origins remain poorly understood.

Sedimentary Rocks

Sedimentary rocks typically form in layers. The layers originate as loose sediment, most commonly deposited underwater. They harden into rock as they become deeply buried by more sediment.

Most sedimentary rocks in Nova Scotia are made of broken-up, weathered fragments eroded from older rocks. The particle size of the sediment determines the rock type: Tiny clay particles form mudstone or shale; sand forms sandstone (or, if its grains are very small, siltstone); and gravel and larger fragments form conglomerate.

On continental margins and in deep ocean basins, turbidites are common (sites 17–24). They are deposited during specific events, sometimes triggered by earthquakes, in which a sediment-laden current rushes down a steep underwater slope. In shallower areas along the coast (sites 5, 34), sediment is distributed by tides and currents. Typically, grain size decreases farther from the shore.

On land, rivers move sediment from highlands into lower-lying areas. Steep slopes, for example beside active faults, create energetic flows of water (sites 32, 33, 38) that can transport gravel-sized sediment or even boulders. In areas of gentler slope, rivers deposit sand, silt, and mud in the river bed and, during floods, in adjacent wetlands (sites 35, 39, 40). If abundant plant material accumulates, burial by younger sediment may transform the plant remains into coal (site 40).

In hot, dry continental regions with infrequent rain, rivers often run dry but may flood after violent or prolonged storms. Sedimentary rocks formed in this setting tend to have a characteristic red or orange colour and are known as continental red beds (sites 41–44, 48). The individual particles of sediment are not red. Instead, as they are deposited and buried under new layers of sediment, each particle becomes coated with colourful ferric iron oxide carried by circulating groundwater.

Sandstone red beds, Five Islands (site 48).

Iron oxide takes two forms, depending on environmental conditions. Reddish ferric oxide (Fe_2O_3) forms in oxygen-rich environments with neutral pH (neither acid nor base). When less oxygen is available or pH is altered, dark grey ferrous oxide (FeO) forms instead. The decay of plants, algae, or other biological matter uses a lot of oxygen and can result in grey layers or patches in an otherwise red or orange sedimentary rock (sites 42–44, 48).

Some sedimentary rocks do not contain weathered pieces of older rock, but instead form chemically when minerals precipitate directly from sea or lake water or with the help of organisms such as corals. Carbonate rocks like limestone form in this way (sites 10, 37). When sea water evaporates in extreme conditions—for example in a shallow tropical sea—the water becomes more and more salty. Eventually, salt crystals accumulate on the sea floor. Minerals and rocks formed by this process are known as evaporites (site 37). They include gypsum (calcium sulphate + water), anhydrite (calcium sulphate), and halite (sodium chloride).

Metamorphic Rocks

Metamorphic rocks form when existing rocks of any type recrystallize due to changes in temperature and/or pressure—for example, when a region of the Earth's crust is buried under a colliding tectonic plate. Metamorphic rocks are classified based on the intensity, or grade, of the metamorphism. The different grades are defined based on the presence of specific metamorphic minerals.

Phyllite, Pillar Rock (site 13).

Carboniferous and younger rocks in Nova Scotia have never been deeply buried, so they have been little affected by metamorphism. In contrast, Nova Scotia's oldest rocks have been very deep in the Earth—30 to 40 kilometres below the surface (site 1). Rocks that originated as muddy sediment are especially revealing of metamorphic grade. As pressure and temperature increase, the minerals recrystallize, transforming mudstone into slate, phyllite, schist, and finally gneiss (sites 13, 14).

In parts of southern Nova Scotia, rocks were not buried so deeply, but they were subjected to high temperatures. They contain minerals that form at high temperature but relatively low pressure, such as andalusite, garnet, staurolite, and cordierite (sites 26, 27). Under some high-grade conditions, migmatite may form when dark and light minerals separate into contrasting layers, with some minerals even melting (sites 1, 2, 28).

Folds and Faults

During tectonic plate interactions, rocks can be folded (sites 23, 28, 42) as crustal blocks move toward or slide past one another. When this happens, flaky metamorphic minerals like mica may become aligned, resulting in foliated rock types that split apart easily (sites 13, 14, 20, 23, 24). Where folds have formed on a very large scale, their whole shape may not be visible in a single outcrop, and the only indication of folding is that the rock layers are tilted.

Folds in a fault zone, Wharton (site 31).

Intense deformation during fault movements can create characteristic broken and sheared rock textures such as breccia or mylonite (sites 2, 31, 48). Intense deformation can also open pathways for mineral-rich fluids. For that reason, folds or faults may control the location of valuable mineral deposits (sites 21, 31, 36).

Fault movements also play an important role in shaping the landscape, in places producing dramatic, linear breaks in topography (sites 31, 32) or steep-sided valleys (sites 43, 44). Because of this effect on topography, faults often significantly influence where and how sedimentary rocks form.

Exploring Further

Canadian Federation of Earth Sciences. *Four Billion Years and Counting*, Ch. 1, "On the Rocks" (pp. 4–18).

Bishop, Arthur C., Alan R. Wooley, and William R. Hamilton. *Guide to Minerals, Rocks & Fossils*. Firefly Books Ltd., 2005.

Plate Tectonics

The story of Nova Scotia's geological past is in large part a story of plate tectonics. In this section you'll find some of the ideas and terms geologists use when thinking and writing about tectonic plates.

Tectonic Plates

Surrounding the Earth's hot, molten outer core is the mantle. Solid but capable of flowing, the mantle carries heat toward the surface of the planet by circulating slowly. Hot regions of the mantle rise toward the Earth's surface and cooler regions of the mantle sink back toward the core in a process known as mantle convection.

The Earth's crust (including all the continents and ocean basins) plus the cooler, upper part of the mantle form a thin outer layer called the lithosphere. Since at least 2,500 million years ago, the lithosphere has been rigid enough to form well-defined tectonic plates, which are about 100 kilometres thick. These plates are always moving, related in a complex way to the flow of hot mantle material beneath them.

Tectonic plates move relative to one another at rates of several centimetres per year. Plate motion causes earthquakes and volcanic activity along plate boundaries, making present-day boundaries easy to recognize. That same motion causes continents to "drift" relative to one another and, occasionally, collide. They are not really "drifting," but are instead carried along as part of a larger tectonic plate.

Plate Interactions

Three types of plate motion describe all plate interactions: divergent, convergent, and transform. For plate motion occurring at the present day, distinctive patterns of earthquake and volcanic activity reveal the boundary type. But what about plate motions in the past? Igneous, sedimentary, and metamorphic rocks provide clues to plate interactions that took place millions of years ago, because plate interactions led to the environments in which the rocks formed and changed. Each type of plate interaction leaves a "signature" of rock characteristics.

Divergent Boundaries

At divergent plate boundaries, hot regions of the mantle flow upward and then away from the plate boundary, moving the plates apart. The Earth's crust breaks, causing shallow earthquakes. Molten rock rises through the cracks, forming intrusions below the surface and volcanic eruptions above.

When divergent motion affects an area of continental crust, continental rifting may lead to the formation of an ocean basin. In Nova Scotia, sedimentary and volcanic rocks associated with continental rifting to form the present-day Atlantic Ocean are well displayed (sites 43–48).

Early in the process, dykes may mark the site of crustal breaks (site 47). As rifting continues, a steep-sided rift valley forms and deepens over time as the lithosphere beneath the valley floor becomes thinner. Eventually the continental crust may separate completely, the area drops below sea level, and a narrow zone of ocean crust develops. Some rifts never develop into ocean basins. Their history is recorded by the sedimentary rock layers that filled in the rift valley over time (sites 43, 44).

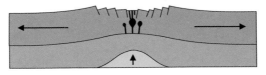

Early Stage

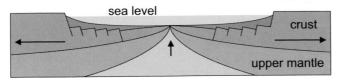

Later Stage

Divergent boundaries.

If the plates continue to move apart, the ocean basin widens. Along the shores of the separated continents, the steep valley walls formed by rifting are replaced over time by stable continental margins known as passive margins. In Nova Scotia, sedimentary rocks formed at two different and unrelated passive margins are preserved, one Ediacaran (site 10) and one early Paleozoic (sites 17–23). A third, present-day passive margin exists offshore.

Mid-ocean ridges mark divergent plate boundaries within areas of ocean crust. To date, no remnants of ocean crust have been found along ancient plate boundaries in Nova Scotia, presumably because it was all carried back into the Earth via subduction along convergent boundaries.

Convergent Boundaries

At convergent plate boundaries, cool regions of the mantle flow toward the plate boundary, then down into the Earth. The plates move toward one another. Convergent boundaries cause deep-seated, severe earthquakes and the Earth's most extreme topography in the form of ocean trenches and mountain ranges.

New convergent boundaries form within ocean crust, usually in older, cooler regions away from the spreading centre. The plate tears apart as one edge sinks, forming a subduction zone. Once established, the subduction zone carries one of the two plates down into the mantle. A deep ocean trench forms along the plate boundary. As the ocean crust descends, it loses water that had accumulated during its interactions with the ocean. The water moves into the overlying mantle, causing it to melt. The resulting magma moves upward through the lithosphere and erupts at the surface (sites 8, 9, 25).

If this happens within ocean crust, the volcanoes typically form an arc-shaped chain of islands (island arc) above the subducting plate. If continental crust exists along a convergent boundary, the volcanoes erupt on land, and they form a magmatic arc of volcanoes and igneous intrusions along the continental margin.

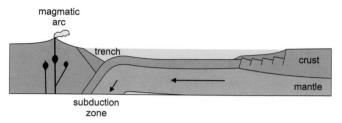

magmatic arc

trench

crust

mantle

subduction zone

Convergent boundary.

If ocean crust and continental crust converge over millions of years, islands, microcontinents, and other landmasses in the ocean basin will collide with the continent in a process of crustal accretion. Ultimately the ocean basin is destroyed, and the continents once separated by the ocean collide. One block of continental crust may ride partly over the other, but continental crust cannot sink into the mantle—it is too buoyant. Instead, continental crust in the collision zone becomes thicker. Lower regions of the thickened crust become hot and are metamorphosed (sites 13, 14, 26–28). If they become hot enough (700° to 800° or more), they partially melt, generating magma that rises through the crust to form igneous intrusions (sites 29, 30).

Transform Boundaries

At transform boundaries, tectonic plates slide past one another. Crust is neither formed, as at ridges, nor destroyed, as in subduction zones. Earthquakes occur as shallower regions of the crust break along faults. At deeper, hotter depths, transform motion distorts the Earth's crust to form rocks such as mylonite (site 2). Transform motion can rupture the crust in complicated patterns, forming deep basins between areas of uplift along faults (sites 32, 33) or prompting igneous activity (site 36).

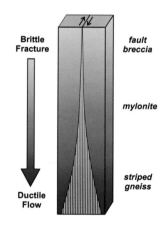

Brittle Fracture

Ductile Flow

fault breccia

mylonite

striped gneiss

Transform boundary.

If transform motion occurs over a long period of time, crustal blocks once adjacent to one another can become widely separated. Some portions of Atlantic Canada appear to have been displaced in this way (sites 3, 31).

Terranes and Microcontinents

Plate interactions can result in small pieces of continental crust, either by rifting of a larger continent to form a microcontinent or by generation of new continental crust in the form of island arcs. Microcontinents and island arcs change position on the globe just as the major continents do. They have been involved in continental collisions in the geologic past and are now embedded in the Earth's mountain belts, including the Appalachians.

When geologists recognize a limited area of continental crust with distinctive geological features and history, they use the word "terrane" to identify and talk about it. Typically, the boundaries between terranes are major faults marking the site of their collision.

Sedimentary rock formations deposited across terrane boundaries, or containing fragments derived from both adjacent terranes, provide clues about when the collision occurred. So do igneous intrusions that cut across terrane boundaries. Nova Scotia includes several distinct terranes, which were assembled by about 360 million years ago.

The Wilson Cycle

What is the outcome of all this plate motion during vast periods of geologic time? Ocean crust is constantly being created and destroyed, while continents slowly grow in size.

An understanding of plate tectonics led to the realization that throughout Earth history a series of ocean basins opened and closed, causing periodic continental collisions of which mountain belts have been the aftermath. This concept, now widely accepted, is known as the Wilson Cycle after its author, Canadian geophysicist J. Tuzo Wilson. The Wilson Cycle describes the "life cycle" of an ocean basin. Each cycle ends in continental collision, forming a new mountain range.

Exploring Further

Canadian Federation of Earth Sciences. *Four Billion Years and Counting*, Ch. 2, "Dance of the Continents" (pp. 24–37).

Kious, W. Jacquelyne and Robert I. Tilling. *This Dynamic Earth: The Story of Plate Tectonics* (Online Edition). U.S. Geological Survey website, pubs.usgs.gov/gip/dynamic. (An excellent summary with numerous illustrations, it can be viewed online or downloaded as a PDF.)

Smithsonian Institution. This Dynamic Planet—World Map of Volcanoes, Earthquakes, Impact Craters, and Plate Tectonics (Interactive Map). www.volcano.si.edu/tdpmap.

Appalachian and Caledonian Mountain Belts

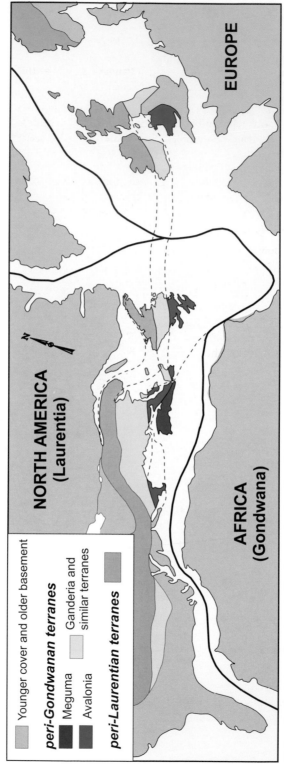

Remnants of microcontinents and other fragments of continental and ocean crust are preserved in the Appalachian mountain belt and its European continuation, the Caledonian mountain belt. They are shown here on a map of present-day landmasses as they were arranged in the supercontinent Pangaea, prior to the opening of the Atlantic Ocean. Dotted lines join disparate portions of the terranes; dark solid lines show where the Atlantic Ocean later opened. (Note: To make continental outlines recognizable, continental shelves are shown here as ocean; however, at the time the mountain belts formed, the entire map area was solid land.)

The Nova Scotia Story

The geological history of Nova Scotia is part of two larger narratives—the formation of the Appalachian mountains and the opening of the Atlantic Ocean. Along the way, the supercontinents Rodinia and Pangaea as well as the now-vanished Iapetus and Rheic oceans played a role. Here's how.

The story begins with the ancient supercontinent of Rodina, formed by a series of continental collisions about 1,100 million years ago. Through the middle of the supercontinent ran a mountain belt (shaded pink in the map below), a part of which is now preserved as the Grenville province of the Canadian Shield. About 750 million years ago, Rodinia began to split apart. Its fragmentation continued for more than 150 million years. A rift valley formed, and as plate motion continued to pull Rodinia apart, the valley deepened. Eventually the continental crust underlying the valley collapsed, and new basaltic ocean crust began to form in the widening gap. The Iapetus Ocean basin was born.

On either side of the new ocean lay Rodinia's diverging parts—two new continents known as Laurentia and Gondwana. The split between Laurentia and Gondwana was along a jagged line, leaving promontories and embayments, or recesses, in Laurentia's margin. The northernmost part of Cape Breton Island was located on one such promontory, which jutted out into the Iapetus Ocean.

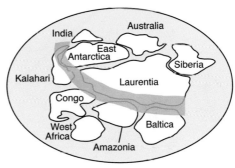

The supercontinent Rodinia.

For about 60 million years, the ocean continued to open. The term "Iapetus Ocean" suggests it was a single, simple ocean basin. But the real story is more complicated. Iapetus was probably similar to areas of the western and southwestern Pacific Ocean today. There, the main Pacific Ocean plate has interacted with the Asian and Australian plates, forming volcanic arcs and small ocean basins of several ages, as well as numerous subduction zones, spreading centres, and microcontinents such as New Zealand and Japan.

Similarly, "Gondwana" was not a single coherent continental mass, but rather a complex region, analogous to present-day Asia. "Asia" is used to refer to an entire region that includes large island arcs such as Japan and Indonesia. "Africa" includes Madagascar, although geologically many consider it a separate microcontinent. In a similar way, the name "Gondwana" is used in this field guide to mean a complex region including islands and microcontinents as well as larger regions of old, stable continental crust. The margin of Laurentia was also complex. Both regions evolved continuously as the Iapetus Ocean opened and closed.

Over time, pieces on or near Gondwana's margin broke away and moved across the Iapetus Ocean to collide with the margin of Laurentia. Their movement was driven by sea-floor spreading along ocean ridges that formed between them and Gondwana.

As the Iapetus Ocean grew smaller, that new ocean, known as the Rheic Ocean, of course grew larger. Closure of the main Iapetus Ocean basin was followed by closure of smaller seaways and basins in the Rheic Ocean, involving the volcanic arcs and microcontinents in complicated plate tectonic interactions.

The schematic map at right shows where these fragments of crust—known as terranes—are preserved in Nova Scotia. Laurentia's continental margin (Blair River Inlier) was separated from Ganderia (NW Cape Breton Island) by the Iapetus Ocean. Scattered across the Rheic Ocean were parts of Avalonia (southeastern Cape Breton Island, northern mainland), and the microcontinent Meguma (southern mainland).

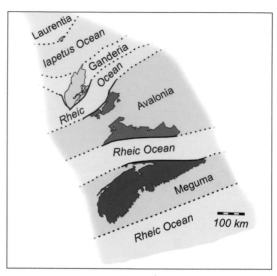

Appalachian terranes of Nova Scotia.

One after another, the pieces were added to Laurentia, each addition causing rocks to be folded, faulted, and metamorphosed, resulting in the intrusion of plutons and batholiths. Each of the pieces that arrived from Gondwana has been given a name—Ganderia, Avalonia, and Meguma—and the mountain-building events associated with their arrivals are known as the Salinic, Acadian, and Neoacadian orogenies, respectively.

During and following the arrival of Meguma, transcurrent plate motion wrenched and distorted the margin between Meguma and Avalonia. Steep-sided valleys formed among the young mountains, which were rapidly eroding due to uplift of the thickened crust.

Finally between about 340 and 300 million years ago, Gondwana's main landmass arrived and collided with Laurentia, of which Ganderia, Avalonia, and Meguma were by then a part. The collision caused a mountain-building event known as the Alleghenian orogeny. With this event, the Earth's landmasses were once again assembled into a supercontinent, this one named Pangaea.

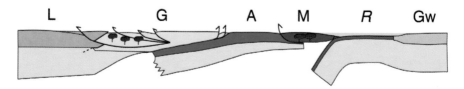

Cross-section: The former edge of Laurentia (L) lies far inland, due to the successive collisions of Ganderia (G), Avalonia (A), and Meguma (M). The Rheic Ocean (R) closed as Gondwana (Gw) approached about 320 million years ago. Red areas are granite intrusions.

What is now Nova Scotia lay right in the middle of Pangaea, near the Earth's equator. Early in Pangaea's history, an arm of the sea flooded large parts of the region. More typically it was covered by rivers, lakes, and swamps, in which both plant and animal life thrived in a humid tropical climate.

The assembly of Pangaea was followed by rapid climate change that led to widespread desert conditions across the supercontinent. Then beginning about 200 million years ago, the central part of the supercontinent began to rift apart. In an early phase of rifting, a series of deep valleys formed west of the present-day Atlantic, represented in Nova Scotia by the Fundy Basin. The early rifts failed, though, and were replaced by a series of interconnecting fractures that would become the present-day Atlantic.

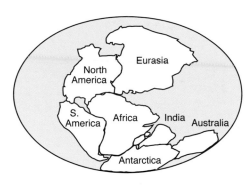

The supercontinent Pangaea.

The new ocean formed approximately, but not exactly, where the Iapetus and Rheic oceans had been. For that reason, parts of Laurentia and Gondwana are found on both sides of the Atlantic Ocean today.

Exploring Further

For a series of maps showing the movement of continental crust over geologic time, visit the PALEOMAP Project by Chris Scotese, www.scotese.com/earth.htm.

Resources

Museums & Interpretive Centres

Nova Scotia offers a variety of museums and interpretive centres with geology-themed exhibits, including:

Cape Breton Miners' Museum, 17 Museum Street, Glace Bay (www.minersmuseum.com).

Creamery Square Heritage Centre, 225 Main Street, Tatamagouche (www.creamerysquare.ca).

Fundy Geological Museum, 162 Two Islands Road, Parrsboro (fundygeological.novascotia.ca).

Joggins Fossil Centre, 30 Main Street, Joggins (jogginsfossilcliffs.net).

Malagash Salt Mine Museum, 1926 North Shore Road, Malagash.

Moose River Gold Mines Museum, 6990 Moose River Road, Tangier; and Moose River Gold Mines Provincial Park (www.novascotiaparks.ca/brochures/Moose_River.pdf).

Museum of Industry, 147 North Foord Street, Stellarton (museumofindustry.novascotia.ca).

Museum of Natural History, 1747 Summer Street, Halifax (naturalhistory.novascotia.ca).

Sydney Mines Fossil Centre, 159 Legatto Street, Sydney Mines (sydneyminesheritage.ca).

Books

Allaby, Michael. *A Dictionary of Earth Sciences*. Oxford University Press, 2008.

Atlantic Geoscience Society. *The Last Billion Years: A Geological History of the Maritime Provinces of Canada*. Nimbus Publishing, 2001.

Barrett, Clarence. *Cape Breton Highlands National Park: A Park Lover's Companion*. Breton Books, 2014.

Canadian Federation of Earth Sciences. *Four Billion Years and Counting: Canada's Geological Heritage*. Nimbus Publishing, 2014. (Illustrations from the book are available online at www.fbycbook.com.)

Websites

Maps

Atlas of Canada, atlas.gc.ca/site/english/toporama.

Natural Resources Canada GeoGratis portal (free downloadable geospatial data), geogratis.cgdi.gc.ca.

Nova Scotia parks and protected areas, www.novascotia.ca/parksandprotectedareas/plan/interactive-map.

Geology

Nova Scotia Department of Natural Resources, Geoheritage Resources, novascotia.ca/natr/meb/geoheritage-resources.

Nova Scotia Department of Natural Resources: Maps, Reports, and Data, novascotia.ca/natr/meb/maps.

Nova Scotia Special Places Protection Act, nslegislature.ca/legc/statutes/specplac.htm.

Museum of Nova Scotia, museum.novascotia.ca/about-nsm/about-heritage/special-places-protection-act.

Smithsonian Institution, Geologic Time: The Story of a Changing Earth, paleobiology.si.edu/geotime/main/index.html.

Touring and Hiking

Canadian Hydrographic Service (tide tables), www.tides.gc.ca.

Leave No Trace, www.leavenotrace.ca.

Nova Scotia atlas, www.novascotia.ca/snsmr/placenames.

Nova Scotia parks, www.novascotiaparks.ca.

Nova Scotia trails, www.trails.gov.ns.ca.

Nova Scotia webcams (real-time local conditions), www.novascotiawebcams.com.

TrailPeak (trail maps and information), trailpeak.com.

Trip Planner

These two pages provide an easy way to plan your geological excursions. The trip planner lists the book's 48 sites—as well as several related museums and interpretive centres—in geographic order. The list begins near Halifax, proceeds west along the Lighthouse Route, and circles the province clockwise, encompassing the northern mainland and Cape Breton Island before approaching Halifax again from the east along Marine Drive. Sites can all be accessed from the province's Scenic Travelways as indicated in the Trip Planner matrix below. For a given region, the sites are listed in a logical travel order, but of course many variations are possible.

For more information about the province's Scenic Travelways, visit Nova Scotia's tourism website at www.novascotia.com or consult your copy of the province's current Doers and Dreamers Travel Guide.

No.	Site	NS Scenic Travelway	Hwy.	Hike	Fac.	Rocks	Nearby
20	Rainbow Haven	Halifax Metro	207	1.4	P	s	
22	Point Pleasant	Halifax Metro	3	0.9	P	s	21
29	Peggys Cove	Lighthouse Route	333	0.5		i	
23	Blue Rocks	Lighthouse Route	332	0.1		s	
21	The Ovens	Lighthouse Route	332	0.2	P	s	
19	Green Bay	Lighthouse Route	331	0		s	
47	Cherry Hill	Lighthouse Route	331	0.1		i	28
27	Sandy Point	Lighthouse Route	3	0.1	P	m	
28	The Hawk	Lighthouse Route	330	0.4	P	m	
26	Pubnico Point	Lighthouse Route	335	1		m	
25	Cape Forchu	Evangeline Trail	304	0.8	P	i	
17	Bartletts Beach	Evangeline Trail	1	0.4		s	
24	Cape St. Marys	Evangeline Trail	1	0.5		i, s	
46	Point Prim	Evangeline Trail	303	0.1	P	i	
45	Halls Harbour	Evangeline Trail	359	1.8		i	
44	Cape Blomidon	Evangeline Trail	221	0.3	P	s	
34	Blue Beach	Evangeline Trail	1	1		s, f	37
42	Rainy Cove	Glooscap Trail	215	1.3		s, f	
43	Burncoat Hd	Glooscap Trail	215	0.3	P	s	
35	Victoria Park, Truro	Glooscap Trail	2	1.5	P	s	
2	Economy River Falls	Glooscap Trail	2	2.1	P	i, m	
48	Five Islands	Glooscap Trail	2	0.9	P	i, s, m	
48	Fundy Geological Museum	Glooscap Trail	2	0	M		
31	Wharton	Glooscap Trail	209	0		m	46
39	Joggins Fossil Centre	Glooscap Trail	242	0	M		
39	Joggins Cliffs	Glooscap Trail	242	0.8	P	s, f	
37	Malagash Salt Mine Museum	Sunrise Trail	6	0	M		
41	Creamery Square Heritage Ctr.	Sunrise Trail	6	0	M		

Site Name and Number (as found in the Table of Contents and site headings)

Scenic Travelway (the provincial Scenic Travelway on which the site is located)

Nearest Highway (the highway pictured in the road map for the site)

Hiking Distance (the round-trip distance between the parking location and the main outcrop, in kilometres)

Facilities (P, located in a national, provincial, or community park, wilderness, or other protected area; M, the listed site is a museum or interpretive centre)

Rock Types (i, igneous; s, sedimentary; f, fossils; m, metamorphic)

Nearby Points of Interest (see Related Outcrops section of the listed site for details)

No.	Site	NS Scenic Travelway	Hwy.	Hike	Fac.	Rocks	Nearby
38	Balmoral Grist Mill	Sunrise Trail	256	0.3		s, f	37
41	Cape John	Sunrise Trail	6	0.1	P	s	
40	Museum of Industry	Sunrise Trail	374	0	M		
5	Arisaig Provincial Park	Sunrise Trail	245	1	P	s, f	
4	Arisaig Point	Sunrise Trail	245	0.3		i	
3	Cape Porcupine	Ceilidh Trail	104	0	P	i	
37	Finlay Point	Ceilidh Trail	19	0.3		s, f	
16	Grand Falaise	Cabot Trail	CT	0	P	i	39
13	Pillar Rock	Cabot Trail	CT	0.4	P	s, m	
1	The Lone Shieling	Cabot Trail	CT	0.5	P	i, m	
33	Deulach Ban Falls	Cabot Trail	CT	0.1	P	s	37
15	Green Cove	Cabot Trail	CT	0.2	P	i	
14	Ingonish Wharf	Cabot Trail	CT	0.1		i, s, m	11
12	Cape Smokey	Cabot Trail	CT	0.3	P	i	
11	North River	Cabot Trail	CT	1.6	P	i	10, 37
40	Sydney Mines Fossil Centre	Bras d'Or Lakes	305	0	M		
7	Coxheath Hills	Bras d'Or Lakes	125	3.5	P	i	
40	Glace Bay Miners Museum	Marconi Trail	28	0	M		
40	Glace Bay	Marconi Trail	28	1		s	
9	Louisbourg Lighthouse	Fleur de Lys Trail	22	0.5	P	i	
8	Kennington Cove	Fleur de Lys Trail	22	0.5	P	i	
6	Point Michaud	Fleur de Lys Trail	247	0.2	P	i	
36	St. Peters Battery	Bras d'Or Lakes	4	0.6	P	i	
32	Marache Point	Fleur de Lys Trail	206	1.1		s	
10	Marble Mountain	Bras d'Or Lakes	4	0		s, m	
30	Black Duck Cove	Marine Drive	16	2.6	P	i	
18	Taylors Head	Marine Drive	7	1.1	P	s	
21	Moose River Gold Mines Mus.	Marine Drive	7	0	M		

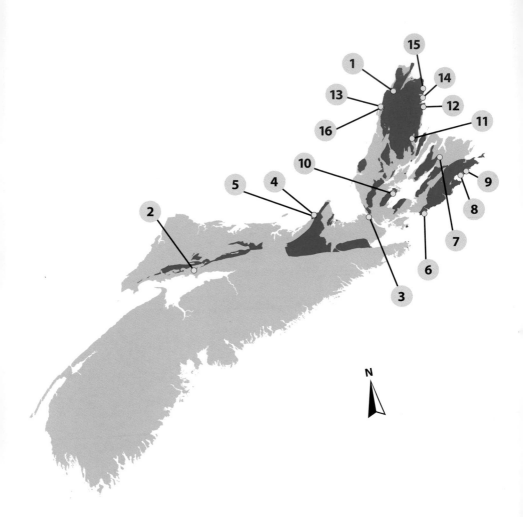

FOUNDATION

At a Glance

Locations

Cape Breton Island; mainland south to the Cobequid-Chedabucto fault system

Origin

Microcontinents colliding with Laurentia

Key Features

Canadian Shield fragment
Volcanic island arcs
Ancient ocean sediments
Granitic intrusions

At these sites, you can …

1	The Lone Shieling	Spot samples of the Canadian Shield.
2	Economy River Falls	Visit the margin of the supercontinent Rodinia.
3	Cape Porcupine	Pinpoint an isolated speck of Avalonia.
4	Arisaig Point	Stand on an Ordovician lava flow.
5	Arisaig Provincial Park	Follow in the footsteps of famous geologists.
6	Point Michaud	Enjoy the surf beside an ancient sea floor.
7	Coxheath Hills	Climb the remains of violent volcanic eruptions.
8	Kennington Cove	Find deposits of fine volcanic ash.
9	Louisbourg Lighthouse	Search for "bombs" from volcanic explosions.
10	Marble Mountain	Bask in the quiet of a shallow ancient sea.
11	North River	Explore the features of a magma chamber.
12	Cape Smokey	View the coastline from atop a unique granite.
13	Pillar Rock	See what metamorphism does to mud.
14	Ingonish Wharf	Take a census of rocks from the highlands.
15	Green Cove	Examine the behaviour of a granitic melt.
16	Grand Falaise	Unscramble a topsy-turvy sequence of events.

Laurentia, Avalonia, and Ganderia: These three landmasses, joined during formation of the Appalachian mountains, were the founding elements of northern Nova Scotia's geological past.

Representing the continent Laurentia is one small corner of the Cape Breton Highlands (site 1) containing remnants of a billion-year-old mountain range. Avalonia (sites 2–9) and Ganderia (sites 10–16) originated as separate regions along the edge of the continent Gondwana. Each eventually split off to become microcontinents that drifted toward and ultimately collided with Laurentia as the Iapetus and Rheic oceans closed.

Economy River (site 2).

Mainland Nova Scotia and southeastern Cape Breton Island preserve separate, distinctive parts of Avalonia. The province's oldest and youngest Avalonian rocks are found on the mainland. Early in its history, about 750 million years ago, a subduction zone caused igneous activity (site 2) along the edge of the supercontinent Rodinia. Granite (site 3) intruded high into the crust provides a glimpse of Avalonia still joined to Gondwana about 600 million years ago. Later, as a separate microcontinent, Avalonia continued its restless ways. Around 450 million years ago, it was the site of volcanic eruptions (site 4). Among its youngest rocks are the remains of a stormy shoreline (site 5) where sediment accumulated about 430 million years ago.

Fourchu Head (site 8).

Pillar Rock (site 13).

Cape Breton Island's distinctive record of Avalonia's history includes a fragment of ocean floor formed near the margin of Rodinia nearly 700 million years ago (site 6). Later, along the margin of Gondwana from 620 to 560 million years ago, much of Avalonia was a region of tall, explosive volcanoes (sites 7, 8, 9).

In Nova Scotia, rocks of Ganderia are restricted to Cape Breton Island. When Ganderia was part of Gondwana's margin more than 600 million years ago, limestone formed there in a warm, shallow sea (site 10). That quiet setting was disrupted about 560 million years ago when a subduction zone formed beneath it, generating great volumes of magma (site 11).

A small volume of granite intruded Ganderia about 495 million years ago (site 12). Meanwhile, belts of sedimentary and volcanic rock accumulated on and beside the microcontinent. Some of them were deeply buried and metamorphosed (sites 13, 14) when Ganderia's northern margin collided with Laurentia about 430 million years ago.

Avalonia collided with Ganderia's southern margin soon afterward. Lower regions of the crust melted, resulting in many granite intrusions and a few basalt eruptions about 375 million years ago (sites 15, 16). Mountain-building in the Appalachians continued with movement on large faults (site 16).

Green Cove (site 15).

29

East of Pleasant Bay, the Lone Shieling and the low wall nearby are made from a variety of locally derived field stones and beach boulders.

Encountering Rodinia
Searching for Nova Scotia's Oldest Rocks

On North Mountain east of Pleasant Bay, the Cabot Trail crosses part of the Blair River Inlier, an area of rocks more than a billion years old. A remnant of a now-eroded mountain range, the inlier once lay at the heart of an ancient supercontinent, Rodinia. You're not allowed to stop and visit the outcrops—it's too dangerous on the steep, winding highway. What to do?

Nearby is the Lone Shieling, erected in 1934 in the Maritime region's largest old-growth hardwood forest. It provides a unique monument to times gone by, including a crofter's cottage typical of the region's early Scottish settlements. Fortunately, the park memorializes Cape Breton Island's ancient geological heritage as well as its Scottish roots.

The walls of the Lone Shieling and the low wall enclosing its grassy surrounds are both made from local stones, including those of the Blair River Inlier. Normally you can't tell the age of a rock just by looking at it. But here at the Lone Shieling you have a good chance of spotting some rare and ancient examples.

Getting There

Driving Directions

Along the Cabot Trail about 6 kilometres east of the community of Pleasant Bay, watch for signs for the Lone Shieling, which is visible in a small grassy area along the south side of the road. As seen on the map, there are related points of interest nearby; see Exploring Further for details.

Where to Park

Parking Location: N46.80984 W60.73300

The entrance to the parking area is on the south side of the highway, west of (downhill from) the structure. Park in one of the designated spaces.

Walking Directions

From the east side of the parking area, follow a short path through the woods and slightly uphill to an open, grassy area beside the highway. The Lone Shieling quickly becomes visible through the trees.

Notes

This portion of the Cabot Trail is a long steep grade; travel toward Pleasant Bay is downhill. The site lies within the boundaries of Cape Breton Highlands National Park and requires a valid park pass.

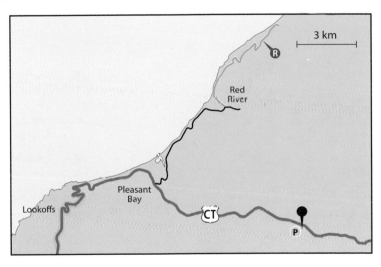

1:50,000 Map

Pleasant Bay 011K15

Provincial Scenic Route

Cabot Trail

On the Outcrop

On the Lone Shieling itself (**a, b**) and in the low wall nearby (**c, d**) are examples from the Blair River Inlier. Rock types include (**a**) pink-grey granulite, (**b**) banded quartz-feldspar gneiss, (**c**) white anorthosite, and (**d**) dark pink syenite.

Outcrop Location: N46.80957 W60.73106

Search the stone walls for boulders from the Blair River Inlier. One variety, called granulite (photo **a**), was metamorphosed under conditions of extreme pressure and temperature deep in the Earth's crust. The granulite has vague, diffuse grey and pink colour variations and an even, sugary texture reflecting the very hot, dry conditions under which it formed.

Beside the doorway of the shieling is a grey and white boulder, a banded gneiss (photo **b**). It began as an igneous rock rich in quartz and feldspar. The banding formed during heating and deformation—any colour variations in the rock were smeared into thin layers.

The very pale grey or white boulders here are anorthosite (photo **c**), made almost entirely of plagioclase feldspar. At least two examples can be found in the low wall along the side that faces the shieling (9 metres and 27 metres from the west end of the wall). Another conspicuous rock type is a dark pink variety known as syenite (photo **d**). A close cousin of granite, this rock is mostly potassium-rich feldspar with almost no quartz.

1000	900	800	700	600	500
Z_1		Z_2		Z_3	$\in$

FYI

- Not all the boulders here are from the Blair River Inlier. Many are younger igneous and metamorphic rocks from the central part of the Cape Breton Highlands farther south.

- Geologists use the word "inlier" to identify a discrete area of older rock surrounded by unrelated younger rock. An inlier may be exposed because it has been uplifted along faults or because it is more resistant to erosion than the surrounding younger rocks.

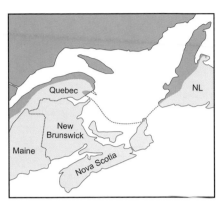

- The Blair River Inlier is part of Laurentia (shaded pink in the map at right). It is preserved on Cape Breton Island because it was located on a bulge, or promontory, of the margin of Laurentia.

Related Outcrops

Rocks of the Blair River Inlier (shaded pink in the map at right) form roadside cliffs along the Cabot Trail between the summit of North Mountain and the community of Pleasant Bay. However, stopping along this steep, winding section of highway is dangerous and prohibited.

West of Pleasant Bay, the Cabot Trail follows a series of switchbacks and s-curves on the slopes of MacKenzie Mountain (see Getting There). In clear weather, several pull-offs along this section of the highway provide views of the steep-sided Blair River Inlier to the northeast.

Exploring Further

For more about rock types including those of the Blair River Inlier, visit the online Imperial College Rock Library at wwwf.imperial.ac.uk/earthscienceandengineering/rocklibrary/index.php.

500   400   300   200   100   0

€ O S D C P Ŧ J K Cz

Related Outcrop

At the Gampo Abbey north of Red River, (**a**) a rock wall surrounding the stupa includes (**b, c**) boulders of anorthosite from the Blair River Inlier.

Outcrop Location: N46.86717 W60.75017

For additional encounters with Nova Scotia's most ancient rocks, from the Cabot Trail in Pleasant Bay, turn (N46.82308 W60.79967) north. Follow Pleasant Bay Loop to Red River Road and continue north to the community of Buddhist monks at Gampo Abbey. Beside Red River Road, north of the main buildings of the abbey, is a stupa (photo **a**)—"a monument that symbolizes the enlightened mind of the Buddha," as explained at the site. A wall frames inspirational plaques around the stupa. Many of the boulders surrounding the plaques come from the Blair River Inlier.

Look for the plaque that says, "In postmeditation, be a child of illusion." Along the right-hand side of the plaque is a sample of snowy white anorthosite streaked with bands of grey (photo **b**). Other, somewhat darker examples of anorthosite can be found nearby (photo **c**).

Near the lower left corner of the same plaque is a darker grey, finely mottled rock representative of the ancient Otter Brook gneiss. Next to it are examples of a dark red syenite characteristic of the nearby inlier. Based on these examples, you can find other similar boulders in the wall.

Otter Brook gneiss.

1000   900   800   700   600   500

Z_1 Z_2 Z_3 €

FYI

- The boulders in the low wall around the stupa were gathered from Pleasant Bay and the mouth of the Red River. The granite of the inscribed stone plaques is not from Nova Scotia. However, the sills under the plaques and the base of the monument are made of Carboniferous sandstone from a quarry near Chéticamp—the same sandstone used to make that town's famous St. Peter's Church.

- Traces of uranium and lead in tiny crystals of zircon from rocks in the Blair River Inlier have been used to determine their age. The inlier's igneous rocks, such as anorthosite and syenite, formed between 1,100 and 980 million years ago, intruding into even older rocks about 1,500 million years old.

- Rocks of this same age are found in many parts of the world, including North America, northern Europe, Africa, and India. They formed as numerous ancient continents collided to create mountain belts during assembly of the supercontinent Rodinia.

- The gneiss, anorthosite, and syenite of the inlier formed deep in the Earth's crust, 30 or 40 kilometres below the surface. Hundreds of millions of years later they were uplifted when the Appalachian mountain belt formed.

From Red River Road the Blair River Inlier appears as a remote, steep-sided headland in the Polletts Cove-Aspy Fault Wilderness Area.

Exploring Further

For more information about Laurentia and the supercontinent Rodinia, see *Four Billion Years and Counting* (Canadian Federation of Earth Sciences, 2014), pp. 93–97.

00   400   300   200   100   0

Є   O   S   D   C   P   Ŧ   J   K   Cz

A very old, strongly layered grey gneiss is exposed in the cliffs and riverbed at Economy River Falls. Areas of lighter pink rock are a much younger granite.

Oldest Avalon
Ancient Gneiss of the Cobequid Highlands

The Earth just can't seem to leave supercontinents in one piece. Take Rodinia, for example. It formed from the collision of numerous continental blocks around 1,100 million years ago (see site 1), but by 750 million years ago new tectonic activity had begun that would lead to Rodinia's breakup.

At Economy River Falls you can visit part of the Earth's crust that formed during that time of transition. The grey gneiss here is part of the oldest known continental crust in Avalonia, from New England to Newfoundland. Originally an igneous rock similar to granite, it intruded into even older rock, fragments of which can also be found along the trail.

Later, but while still deep in the Earth, the rocks were smeared out in a zone of intense deformation, giving them a unique mineral fabric. Today they are located along the much younger Cobequid-Chedabucto fault system, which marks the southern boundary of Avalonia in Nova Scotia. The trail and waterfall are located along the fault scarp, which forms a conspicuous topographic break across the region.

Getting There

Driving Directions

Along Highway 2 in the community of Economy, turn (N45.38482 W63.91441) north onto River Phillip Road. Follow this gravel road for about 7 kilometres, then turn right (east) and continue about 300 metres to a parking area.

If you like, pause on River Phillip Road about 2 kilometres from Highway 2 (N45.40307 W63.91820) to view the trace of the Cobequid fault to the north.

Where to Park

Parking Location: N45.44426 W63.92790

Park in the gravel clearing.

Walking Directions

The trail head is on the east side of the parking area. Walk about 500 metres east to a fork on the right, which leads steeply downhill to the base of the falls about 225 metres from the fork. The trail that continues east along higher ground leads to a footbridge above the head of the falls about 325 metres from the fork.

Notes

The gravel road to Economy River Falls is not well maintained and may include large potholes and minor washouts. The hiking trail is clearly marked but rustic. Puddles and small streams form on it in wet weather. The walking distance given is for the round trip to sites both below and above the falls. This site is protected under provincial law as part of the Economy River Wilderness Area.

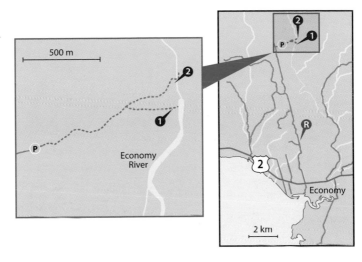

1:50,000 Map

Bass River 011E05

Provincial Scenic Route

Glooscap Trail

On the Outcrop

(**a**) Outcrops of the Economy River gneiss have a layered, mylonitic fabric. (**b**) Worn boulders of the gneiss in the riverbed show the mylonitic fabric in detail. (**c**) Fragments of ironstone show a similar but less obvious fabric.

Outcrop Location: N45.44600 W63.91967

The riverbank below the waterfall is a good place to view the Economy River gneiss. Other examples can be found along the trail to the footbridge and beside the gorge above the falls (for example, N45.44701 W63.91938). You'll see three main types of rock.

The most common rock is the grey gneiss that forms thick layers tilted downstream or slanting across the riverbed. Clean outcrops of the gneiss have a lined appearance, in places resembling a bundle of grey pencils. On water-smoothed boulders you can clearly see the wavy, ribbon-like pattern caused by extreme deformation. Rocks having this fabric are known as mylonite. Interlayered with the grey gneiss are conspicuous pink sheets of younger granite. It has a less extreme mylonitic fabric.

Along the trail or riverbank, also look closely for boulders of very dark grey rock. Some of these feel especially heavy because of their high density. If you happen to have a magnet, you will find that the rock attracts it. Known as ironstone due to its high iron content, this dense rock is older than the gneiss.

1000	900	800	700	600	50
Z_1		Z_2		Z_3	€

FYI

- Mylonite forms in fault zones deep in the Earth's crust where rocks are hot and ductile. Faults cause the crust to fracture at shallow levels, but at depth a wide zone of smeared-out rock is formed during slow, continuous movement.

- The pink granite intruded the grey gneiss about 360 million years ago while the mylonitic fabric was forming.

- Traces of uranium and lead in zircon grains from the grey gneiss show that it formed about 734 million years ago. Similar rocks farther east in the Cobequid Highlands have an even older age of 750 million years.

Pink granite.

- The magnetic ironstone at Economy River Falls is part of an older sequence of sedimentary and volcanic rock layers that may have formed in a rift as Rodinia broke apart. These ancient rocks have been found to contain grains of zircon eroded from an unknown source at least 950 million years old.

- The ironstone is a sedimentary rock. It may have formed when volcanic activity in the ocean created an iron-rich brine, leading to the precipitation of iron oxide.

Related Outcrops

In Nova Scotia, rocks of Avalonia appear in three regions (shaded green in the map at right). This site is part of the Cobequid Highlands (C). Farther east are the Antigonish Highlands (A; see sites 3–5) and the Mira terrane (M; see sites 6–9).

The Cobequid fault scarp that forms the southern boundary of Avalonia is obvious along a stretch of Highway 209 east of Parrsboro (see site 31) and along Highway 16 southeast of Guysborough (see site 30).

The tall cliffs of Cape Porcupine rise steeply beside the Canso Strait as seen from Port Hastings, Cape Breton Island.

Granite Lineage
A Lone but Useful Fragment of Avalonia

Next time you travel the Canso Causeway, tip your hat to Cape Porcupine looming beside the water. More than 9 million tonnes of the cape's brown granite were piled into the Strait of Canso to create the road link to Cape Breton Island. Its busy quarry is still one of Canada's top producers of industrial aggregate.

For many years geologists thought of Cape Porcupine as a sort of rocky orphan. It's a little speck of Avalonia—barely 4 square kilometres—bordered all around by faults and younger sedimentary layers. The cape itself includes several types of igneous and metamorphic rock. You can see the colour variations in the boulders of the causeway.

As it turns out, Cape Porcupine has some close relatives after all. Its brown granite formed 610 million years ago, as did many others in Avalonia. Younger igneous rocks here match a 480-million-year-old "sister" complex in the Antigonish Highlands. There's no match to Cape Breton Island, though: The Strait of Canso runs along a major fault, and the island once lay 100 kilometres north of Cape Porcupine.

Getting There

Driving Directions

At the eastern end of the Canso Causeway along Highway 104 a swing bridge crosses the Canso Canal. East of the bridge, beside the canal, is a small park with a grassy lawn and a kiosk. About 50 metres east of the bridge, turn northwest into the park's short access road.

Where to Park

Parking Location: N45.64785 W61.41204

Park in one of the designated spaces near the entrance to the park.

Walking Directions

A sidewalk from the parking area to the kiosk leads past a sample of the Cape Porcupine granite and to views of Cape Porcupine across the Strait of Canso.

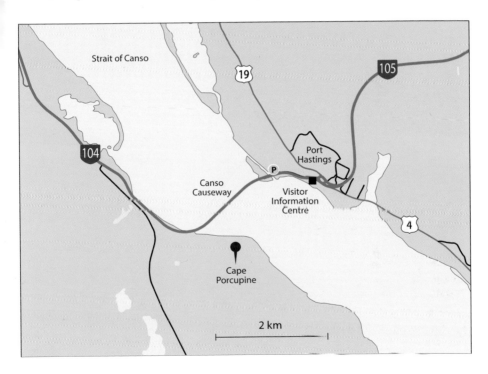

1:50,000 Map

Port Hawkesbury 011F11

Provincial Scenic Route

Ceilidh Trail

On the Outcrop

Aggregate from the quarry at Cape Porcupine is pushed over the cliff and accumulates for loading below. A slightly darker dyke (above arrow at left) forms a natural chute in the brown granite cliff. A smaller dyke (above arrow at right) is also visible.

Outcrop Location: N45.64788 W61.41233

For this site the outcrop location waypoint marks the best place to view the Cape Porcupine Complex—the small park by the Canso Canal in Port Hastings. From there you have a clear view across the strait. The brown cliffs of Cape Porcupine granite are cut by several dark grey dykes.

In the park a large piece of the granite serves as the base for a bronze plaque commemorating construction of the canal. The Cape Porcupine granite is mostly made of orange or brown feldspar, which accounts for the rock's colour. It has less quartz than a typical granite and very few dark minerals, so the colour is quite uniform.

The granite was intruded into the crust at a high level (near the surface), so it has little voids where gas bubbles formed in the thick, sticky magma. On the top surface of the boulder you'll see a finer-grained, darker layer representing

Cape Porcupine granite.

one of the other rock types in the quarry's complex igneous assemblage.

FYI

- The quarry on Cape Porcupine covers more than 300 hectares and has reserves of more than 270 million tonnes of high-quality granite. Its direct access to an ice-free, deepwater port allows year-round shipping of aggregate from the site.

Gazebo by the Canso Canal.

- The Canso fault extends for 100 kilometres beginning offshore near Prince Edward Island. It appears to be truncated by the younger Cobequid-Chedabucto fault near the community of Canso on the mainland. When active long ago, the Canso fault was similar to the present-day San Andreas fault in California, where two crustal blocks are sliding past one another.

- The rocks of Avalonia on the mainland (light green in the map below) are distinct from those of New Brunswick and Cape Breton Island (dark green), which were once side by side. The crustal block east of the Canso fault slid south relative to mainland Nova Scotia. This left contrasting parts of Avalonia on either side of the strait.

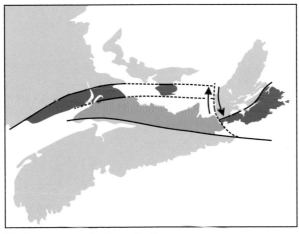

Displacement on the Canso fault.

Related Outcrops

Access to Neoproterozoic rock outcrops in the Antigonish Highlands part of Avalonia is limited. However, along the coast at Arisaig Point and Arisaig Provincial Park (sites 4, 5), Avalonian igneous and sedimentary rocks from the Ordovician and Silurian periods are well exposed.

Pink rhyolite in the foreground and dark green basalt in the background record two types of volcanic eruption at Arisaig Point.

Drifting and Rifting
Bimodal Volcanic Activity in Restless Avalonia

The Ordovician period was a special time for Avalonia. It had broken away from the large continent of Gondwana to the south and was now separated from it by the Rheic Ocean. Laurentia was still far to the north, across the Iapetus Ocean. Avalonia was an independent microcontinent on the move.

In the middle of the Ordovician period, tectonic stresses on the microcontinent caused the lithosphere to pull apart locally and form a rift. Hot, mobile magma rose from the mantle below. Along the way, it heated and partly melted areas of the crust, forming another body of magma, this one thick and sticky with a composition like granite.

The result was a series of bimodal eruptions, that is, volcanic activity producing two types of rock: The basalt (from the mantle) flowed onto the surface as a lava. The rhyolite (from the crust) resulted from explosive nuée ardente ("glowing cloud") eruptions that cooled quickly into colourful, glassy layers. That dramatic interlude is long gone, so now you can visit these contrasting rock types in peace and tranquility beside the lighthouse on Arisaig Point.

Getting There

Driving Directions

On Highway 245 about 6.75 kilometres west of its intersection with Highway 337 at Malignant Cove, watch for St. Margaret of Scotland Catholic Church. Near the church, turn (N45.76025 W62.16238) west onto Arisaig Point Road. You may see signs for the Arisaig lighthouse indicating this route. Follow the road to the public wharf and lighthouse.

Where to Park

Parking Location: N45.76204 W62.17180

Park in the gravel area near the lighthouse.

Walking Directions

From the parking area, walk eastward away from the lighthouse. Footpaths lead across a grassy, rocky knoll. Continue across the knoll and, as conditions allow, cross onto the beach. Outcrops 1 and 2 are located on either side of the gravel flat. Rocks like those at Outcrop 1 are also seen on the knoll.

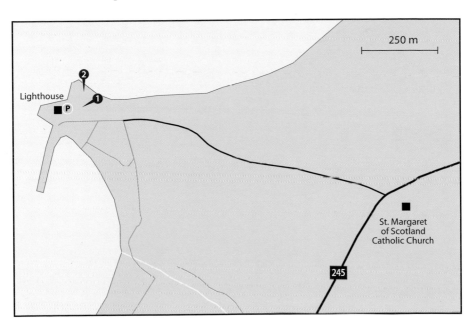

1:50,000 Map

Malignant Cove 011E16

Provincial Scenic Route

Sunrise Trail

45

On the Outcrop (1)

The rhyolite has narrow, multicoloured banding. On some outcrops the banding is folded (see detail below; each square on the black and white scale is 1 centimetre across).

Outcrop Location: N45.76210 W62.17132

The lighthouse at Arisaig Point sits on a layer of rhyolite that extends to the adjacent knoll. The rock is vividly striped in shades of dark red, pink, orange, and grey. The stripes are narrow, only a few millimetres wide. This type of rock is known as a flow-banded rhyolite.

The colourful flow banding has small-scale bends and buckles in some places, but the banding was originally horizontal on average. Much later, the rhyolite layers were tilted as part of a large-scale fold that affected rocks between Moydart Point and Malignant Cove.

The contorted banding formed while the stiff lava was still hot and slowly flowing. It might have taken about a week to cool and stop changing shape. On the little beach, cobbles of rhyolite have been polished smooth and show the banding in beautiful detail.

Banded rhyolite.

1000   900   800   700   600   500

Z₁   Z₂   Z₃   €

FYI

- New Zealand is a microcontinent with similarities to Ordovician Avalonia. Near the middle of New Zealand's north island in the region around Lake Taupo, a similar rift exists with similar bimodal volcanic activity.

- Individual particles in a nuée ardente eruption are so hot that they weld together into a thick, syrupy mass on the ground. In this form, the rock is like molten glass and can flow slowly, contorting the layers. The colour of a layer depends partly on how much the iron in it has been oxidized (more oxidized layers are redder).

- Though their origin was not well understood until recently, the rocks at Arisaig Point have caught the attention of geologists for 150 years or more. In 1870, Rev. David Honeyman described to the Royal Society of London a "porcellanous jasper at Arisaig pier, Frenchman's Barn, &c.," after he encountered the rhyolite in his search for nearby fossils.

Banded rhyolite outcrops dot a knoll behind the Arisaig lighthouse.

Related Outcrops

Bimodal volcanic activity occurs when continental crust is pulled apart. This has happened at various times in Nova Scotia's past. See sites 11 (By the Way) and 24 for additional examples. Much older flow-banded rhyolite occurs in the Coxheath Hills near Sydney (site 7).

Exploring Further

Honeyman, David. "Notes on the Geology of Arisaig, Nova Scotia." *Quarterly Journal of the Geological Society of London*, vol. 26, pt. 1 (1870), p. 491 (available at books.google.ca).

500   400   300   200   100   0

Є O S D C P T J K Cz

On the Outcrop (2)

The basalt exposed at Arisaig Point is dark green with a brown surface that is due to oxidation (rusting) of iron-rich minerals.

Outcrop Location: N45.76243 W62.17130

The basalt at Arisaig Point is on the far side of the small beach, north of the rhyolite. The outcrops are quite fractured, but you can still see some features of the original lava flow.

For example, some parts of the outcrops have a spotted appearance. Rocks with this appearance are said to have an amygdaloidal texture. The lava contained bubbles of volcanic gas as it flowed onto the Earth's surface. The bubbles left voids in the rock as it cooled, and the voids later filled with minerals such as calcite, quartz, or zeolite.

Amygdaloidal basalt.

Like the flow banding in the rhyolite nearby, the amygdaloidal texture of the basalt captures a dramatic moment—in this case, bubbling lava flowing right here—many millions of years ago. Thanks to these freeze-frames provided by nature, anyone can explore exciting volcanic processes.

1000   900   800   700   600   500

Z_1 Z_2 Z_3 ε

FYI

- Since they formed, the rocks at Arisaig Point have not been deformed very much and only slightly metamorphosed, even when Avalonia collided with Laurentia and Meguma. Geologists think this area was located far from the site of those collisions.

- The large, dark boulders placed at Arisaig Point to hinder erosion are gabbro unrelated to the basalt outcrops. Quarried in Georgeville, about 15 kilometres northeast of Arisaig, the gabbro is 605 million years old (similar in age to the Cape Porcupine granite, site 3).

Arisaig lighthouse.

- Paleomagnetic data from this and other sites (diagram below) show that Avalonia was moving north toward Laurentia at the rate of about 5 centimetres per year during the Ordovician and Silurian periods. This site was located at about 40° South at the time the basalt formed.

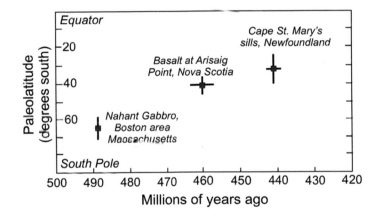

Related Outcrops

The 480-million-year-old igneous rocks at Cape Porcupine (site 3) may have formed in an earlier phase of the same rifting event that caused the bimodal eruptions here.

Exploring Further

For paleogeography maps showing the movement of Avalonia from 500 to 450 million years ago, see *Four Billion Years and Counting* (Canadian Federation of Earth Sciences, 2014), p. 115.

500   400   300   200   100   0

€ O S D C P Ŧ J K Cz

Tilted shale and siltstone layers along the shore in Arisaig Provincial Park are part of an unbroken sedimentary record of the Silurian period.

Stormy Shores
A Much-Studied Record of Silurian Avalonia

As you tread the sand of this little beach in Arisaig Provincial Park, you'll be walking in the footsteps of Canada's early geologists. For more than 175 years the sedimentary rocks here have been studied by pre-eminent naturalists. Visitors in the mid-1800s included Abraham Gesner, David Honeyman, and J. William Dawson.

What attracted these scientific pioneers was a sequence of fossil-bearing sedimentary rocks deposited without interruption throughout the Silurian period (about 440 to 420 million years ago). Measuring nearly 1,400 metres thick, the rock sequence originated as layers of mud and silt beneath shallow ocean waters where a riotous variety of marine organisms lived and died.

The provincial park encompasses just a small part of this famous sequence of sediments. The rock layers not only capture important changes in the fossil record; they also contain clues to environmental conditions on the shores of this part of Avalonia.

On your way to the site, you can visit the park interpretive centre, located in an open pavilion along the trail (N45.75474 W62.17125). There, interpretive panels describe the park's geological history, ancient environments, and fossils.

Getting There

Driving Directions

On Highway 245 about 7.75 kilometres west of its intersection with Highway 337 at Malignant Cove, watch for the entrance to Arisaig Provincial Park. Turn (N45.75468 W62.16648) northwest onto the park road and follow signs to the interpretive centre.

Where to Park

Parking Location: N45.75436 W62.17083

Park in the parking area for the interpretive centre.

Walking Directions

From the parking lot, follow the trail to the interpretive centre—an open, roofed pavilion. From the pavilion a footpath and staircase lead downhill to a series of branching trails. From this point, follow the trail that forks slightly left and leads most directly downhill. It emerges from the woods and follows a seaside cliff top, eventually passing a small cove. On one side of the cove is a lookoff platform and on the other is a stairway providing access to the shore. As conditions allow, walk out of the cove onto the beach and around to the right.

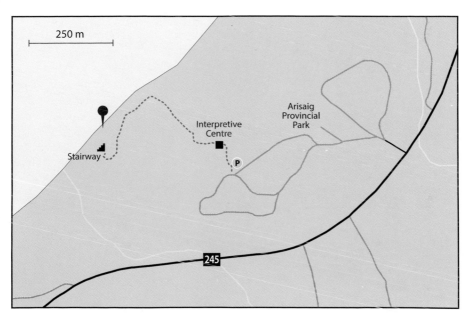

1:50,000 Map

Malignant Cove 011E16

Provincial Scenic Route

Sunrise Trail

On the Outcrop

Outcrops in this part of the park contain thin layers of dark shale and lighter, slightly thicker layers of siltstone.

Outcrop Location: N45.75505 W62.17434

Early in the Silurian period this site lay, as it does now, along a shore periodically swept by storms. The greenish-grey shale layers on this stretch of beach mainly record the influence of those storms (as opposed to the effects of tides or other currents) and are known as tempestites.

As you face the outcrops, the rocks on your left (northeast toward Arisaig lighthouse) are older and those on your right (toward the southwest) are younger. The tops of the layers are on the right-hand side.

Coquina in shale.

Some layers are fissile, breaking apart easily into thin layers. Others are more coherent, and on the top surface of this kind, you may notice some dots or bumps about 0.5 centimetres wide. These are traces of worm burrows.

Between layers of shale you may find pod-like collections of coquina, a rock made of shell fragments. The shells were swept from beaches during a storm and gathered in a channel of fast-moving storm water or similar depression.

FYI

- Farther southwest along the shore, shale becomes less common and siltstone more so. The siltstone is especially rich in fossils, including species of bivalves, brachiopods, crinoids, corals, and graptolites as well as trace fossils such as worm burrows.

- Throughout the Silurian period, the microcontinent Avalonia drifted north toward the equator, drawing closer to Laurentia. The continents eventually collided early during the Devonian period.

- The uninterrupted accumulation of fine-grained sediment here during the Silurian period suggests that the area was tectonically quiet for at least 20 million years. Some geologists interpret this to mean that the site was located on the far side of Avalonia as it approached Laurentia.

A wooden stairway leads down to the shore in Arisaig Provincial Park.

Related Outcrops

Other fossil-bearing Silurian rocks in Nova Scotia were formed on the separate microcontinent of Meguma, which later collided with Avalonia. In Meguma the Silurian period included volcanic activity (see sites 24 and 25).

Exploring Further

To view early drawings of Arisaig fossils, see J. William Dawson's *Acadian Geology* (1855, available at books.google.ca), supplementary chapter, pp. 67–69. For a brief biography of Dawson and others active in the early geological exploration of Nova Scotia, see *The Last Billion Years* (Atlantic Geoscience Society, 2001), pp. 36–38.

For a park brochure and a map of Arisaig Provincial Park's trail system, visit www.novascotiaparks.ca.

A small headland buttresses the northeastern end of the beach at Point Michaud, forming a low cliff of basalt.

Ancient Sea Floor
Oldest Avalonia on Cape Breton Island

Point Michaud offers one of Cape Breton Island's finest beaches—a graceful 2.5-kilometre arc of sand and surf. It's a pristine and peaceful setting, but the rock cliff beside the beach suggests that 670 million years ago, the site was anything but.

Frequent earthquakes may have shaken the ground as nearby, two tectonic plates clashed: One plate slid below the other in a subduction zone. A long, narrow area of weakness formed above the subducted plate, tearing the crust open to form a sort of mini-ocean floor called a back-arc basin. Into the basin poured lava flows of basalt, hissing and steaming underwater. Elsewhere in the basin, mud and fine sand accumulated while super-heated, mineral-rich fluids escaped from below, leaving behind sulphurous deposits of copper, zinc, and other metals.

Welcome to the world of the Stirling belt, an elongated band of igneous and sedimentary rocks that extends northeastward, slicing inland from Point Michaud. Formed along the tectonically active margin of an ancient continent, it is the oldest known part of Avalonia on Cape Breton Island.

Getting There

Driving Directions

From Highway 4 east of St. Peters, turn (N45.65182 W60.85763) south and follow Highway 247 through L'Ardoise. Continue southeast and east along the coast, watching for signs to Point Michaud Beach Provincial Park. As you approach, the long expanse of the beach is visible on the right.

Where to Park

Parking Locations: (Beach) N45.59213 W60.68041
(Picnic area) N45.59228 W60.67819

There is a large parking lot between the road and the beach, and a smaller one by the picnic area. To access the picnic area, fork right from the main road near the beach and follow a gravel road up the hill to a small headland.

Walking Directions

From the beach by the parking lot, there are outcrops in a low cliff about 70 metres to the left (east) as you face the water. They are easily accessed by walking along the beach. Alternatively, from the picnic area, follow the fence east. Near the end of the fence a grassy footpath leads down to a small beach where you can climb onto the outcrops.

Notes

Access to the outcrops requires appropriate tide conditions. The entire beach at this site is a protected area under provincial law.

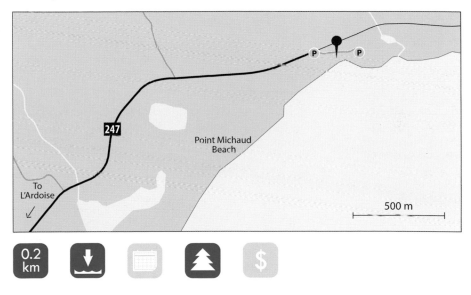

1:50,000 Map

St. Peters 011F10

Provincial Scenic Route

Fleur de Lys Trail

On the Outcrop

The remains of original lava pillows intermingle with the broken pillow fragments and mineral veins of this submarine eruption.

Outcrop Location: N45.59207 W60.67901

As you explore the outcrops below the headland, the first feature you may notice is their markedly green colour. It is due to low-grade metamorphism of the basalt, during which the original, dark green or black igneous mineral, pyroxene, was replaced by brighter green minerals, epidote and chlorite.

Another prominent feature of the basalt is the complex mingling of solid green areas and an irregular web of lighter minerals and fragments. This pattern arose as pillows formed, hardened, fractured, and were pushed aside by new blobs of lava. As the pile of pillows cooled, hot fluids circulated among the pillows and fragments, depositing minerals including quartz and calcite.

Amygdaloidal texture.

Some of the outcrops have a dotted appearance, known as an amygdaloidal texture. The round or oval areas of lighter green contained bubbles that were later filled with minerals deposited from a hot fluid that circulated through the lava.

1000 900 800 700 600 500

Z₁ Z₂ Z₃ €

FYI

- The Stirling belt is named for a deposit of zinc, lead, copper, silver, and gold near Stirling in Richmond County, about 20 kilometres northeast of Point Michaud. Discovered in the 1890s, it is Nova Scotia's only known volcanogenic massive sulphide (VMS) deposit. Its Mindamar mine produced more than 1 million tonnes of ore in the 1930s and 1950s.

- The basalt at Point Michaud has specific chemical characteristics that allow geologists to identify its setting as a back-arc basin. They include relatively low amounts of titanium, iron, and niobium.

- The Stirling belt is about 30 kilometres long but less than 10 kilometres wide. The subduction zone that caused the volcanic activity would have been located parallel to the belt. Some evidence suggests that it was on the southeastern side.

- Few rocks of similar age occur in other regions of Avalonia. One example in the Connaigre Bay region on the south coast of Newfoundland also has similar VMS mineral deposits.

Point Michaud Provincial Park.

Related Outcrops

The extent of the Stirling belt is shaded green in the map at right. On Cape Breton Island, Avalonia also includes younger belts of subduction-related volcanic activity (see sites 7, 8, and 9).

Basalt formed in a variety of other settings and time periods across Nova Scotia. For examples, see sites 4, 36, 45, and 46.

Exploring Further

For more information and videos of hot fluids circulating in the present-day ocean floor, visit the U.S. National Oceanic and Atmospheric Administration online at www.pmel.noaa.gov/eoi/PlumeStudies/BlackSmokers.html.

500	400	300	200	100	0
Є O S	D	C	P R	J	K Cz

The rocky barren atop Coxheath Hills offers excellent views and many outcrops of brick-red volcanic rock known as tuff.

Pyroclastic Display
Explosive Volcanic Eruptions in Avalonia

The Earth's continents grow slowly but continually, almost like living things. Continental material is formed (and some is recycled) in subduction zones. It is added to the existing landmass by volcanoes and by the huge chambers of molten rock that feed them. That happened here.

About 620 million years ago, volcanoes formed in many parts of Avalonia due to one or more subduction zones near the margin of its parent continent. In today's world, they would be the type of volcano that makes headlines, like Mount St. Helens: Widespread destruction. Planes rerouted to avoid engine damage. Spectacular red sunsets for weeks afterward.

As you hike the trails of Coxheath Hills, you are walking across a landscape that was once the scene of many pyroclastic eruptions, including nuées ardentes ("glowing clouds"), thick layers of volcanic ash, and the thud of ejected blocks falling to earth. These days you might hear others' footfalls, or encounter some fog—nothing like the disturbances that accompanied Avalonia's growing pains. The only signs of that ancient turmoil are underfoot.

Getting There

Driving Directions

From Highway 125 near Sydney, take exit 5 for Coxheath and follow Highway 305 south to Coxheath Road. For the trail head, travel southwest on Coxheath Road to Knox United Church in the community of Blacketts Lake. For the access road, watch for a gravel road on the north side of Coxheath Road, near Hanart Wood Drive on the south side.

Where to Park

Parking Locations: (Trail head) N46.07654 W60.29884
(Access road) N46.08153 W60.28166

For the trail head, park in the gravel parking lot beside Knox United Church. For the access road, follow the gravel road around to the right and park near the gate.

Walking Directions

A kiosk with a trail map marks the trail head on the north side of the church parking area. Follow the Main Trail steeply upward. Where the Western Loop forks left, bear right. After about 500 metres, the trail becomes less steep and leads to a series of large outcrops and panoramic lookoffs on the highest ground. The gravel access road is less steep but less scenic than the first section of the Main Trail. It leads to Patterson Trail; follow it and bear left beside the cabin to join the Main Trail near the lookoff.

Notes

These trails are maintained by the Coxheath Hills Wilderness Recreation Association. For more information about the trail system, visit www.coxheathhills.com. The hiking distance given is to the lookoff and back from either parking location.

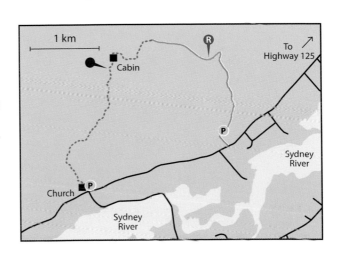

 3.5 km

1:50,000 Map
Sydney 011K01

Provincial Scenic Route
Bras d'Or Lakes Scenic Drive

On the Outcrop

(**a**) Large outcrops provide excellent lookoffs and rock exposures along the Main Trail on Coxheath Hills. (**b, c**) Outcrops worn clean by foot traffic reveal more detail (see text).

Outcrop Location: N46.08725 W60.29598

The dominant rock type along the Main Trail and at the lookoffs on the high ground (photo **a**) at Coxheath Hills has a colour and texture resembling brick. Geologists call this rock a felsic tuff. It is rich in quartz and feldspar, as granite is. But unlike granite, tuff is made of material ejected from a volcano, including glassy ash, crystals, and rock fragments.

In some outcrops, the very fine-grained, dark red matrix is speckled with white crystals 1 to 2 millimetres in size (photo **b**). They are quartz and feldspar crystals that were growing in the molten magma as it slowly cooled beneath the volcano. The red matrix has mineral grains that are too small to see. They crystallized suddenly during the eruption and had little time to grow.

In a couple of large, foot-worn rock pavements near the lookoffs, the speckled pink matrix contains blocks of darker red rock a few centimetres across (photo **c**). These fragments are from slightly older parts of the volcano, broken off and carried along in a later explosion of material.

1000 900 800 700 600 500

Z₁ Z₂ Z₃ Є

FYI

- Volcanic and plutonic rocks the same age as those seen here have been found in other parts of Avalonia, including mainland Nova Scotia, eastern Newfoundland, southern New Brunswick, and southeastern New England. At that time the microcontinent may have resembled modern-day Japan or the Philippines, with many active volcanoes.

- Hugh Fletcher of the Geological Survey of Canada described in his report of 1875 a deposit of copper-rich minerals in the Coxheath Hills. Recent studies suggest that it is a "porphyry" deposit similar to those found in the Andes Mountains of South America. Such deposits form in granite magmas intruded and cooled in the roots of a volcano.

Related Outcrops

On the gravel-paved Access Trail at Coxheath Hills, along the switchback (for example, N46.08802 W60.28330) you will find examples of flow-banded rhyolite. This type of rock forms as the hot, glassy remnants of a nuée ardente slowly sag and ooze along the ground after an eruption (also see site 4).

Flow-banded rhyolite.

Coxheath Hills is part of a larger belt of volcanic rocks and related intrusions (shaded dark and light green, respectively, in the map above) that all formed roughly 620 million years ago. East Bay Hills along the southeastern shore of Bras d'Or Lake is the most extensive example.

Exploring Further

Fletcher, Hugh. "Report of Explorations and Surveys in Cape Breton, Nova Scotia." In *Geological Survey of Canada Report of Progress for 1875–76*, Dawson Bros., 1877, pp. 369–418 (available online at books.google.ca).

500   400   300   200   100   0

€ O S D C P T J K Cz

From the beach at Kennington Cove (far left), it is easy to access clean rock outcrops that emerge from the beach boulders and sand.

Crystal Rain

Ash Deposits as Volcanism Resumes in Avalonia

About 100 years ago, members of a small farming and fishing community made their home around Kennington Cove. Phosa Kinley, a schoolteacher, reported, "Cannon balls brought in from the woods and fields were in almost every house in the district, and several of these were arranged in orderly rows on my desk in the schoolhouse."

Locals had pointed out to Phosa the rock where Brigadier General Wolfe made land in June 1758; she realized the munitions on her desk had rained down during the battle for Louisbourg Fortress. She probably did not realize that the rock on which Wolfe set foot had also rained down from the sky—in the form of volcanic ash.

At least three episodes of plate subduction affected the margin of an ancient continent at the end of the Proterozoic era (see sites 6, 7, and 9). Kennington Cove is part of the youngest such episode. The volcanoes here were like those found today in the Philippines and Indonesia. Their forceful eruptions sent clouds of volcanic ash high into the sky. Hours later, the ash fell here and blanketed the ground.

Getting There

Driving Directions

Highway 22 becomes Main Street in the community of Louisbourg. From the west end of Main Street, follow signs toward the Parks Canada Visitor Centre on Wolfe Street. Travel past the entrance to the Visitor Centre as Wolfe Street becomes Kennington Cove Road and continue southwest. About 8 kilometres from the Visitor Centre, you will pass a parking area and large grassy field along the shore. Continue another 750 metres to a second parking area at the end of the road.

Where to Park

Parking Location: N45.87553 W60.06408

Park in the large gravel parking area at the end of the road.

Walking Directions

From the south side of the parking area, follow a trail through the woods—it emerges into an open, grassy area. Bear left and continue through the grass along the edge of the woods toward the shore. Bear left again and follow another short trail along the wooded shoreline. A set of stairs provides access to the beach. On the beach, turn right and walk about 20 metres to the outcrop.

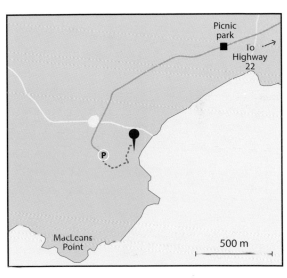

Notes

This site is located within the Fortress of Louisbourg National Historic Site of Canada.

1:50,000 Map

Mira River 011F16

Provincial Scenic Route

Fleur de Lys Trail

On the Outcrop

Subtle colours, tilted layering, and a rough, sandy texture characterize the volcanic rock layers on the beach. Note the darker rock, a dyke, on the right.

Outcrop Location: N45.87568 W60.06250

The rock in Kennington Cove is called a crystal-lithic tuff. It includes abundant visible grains of quartz and feldspar as well as some small rock fragments. The crystals of quartz are especially conspicuous on weathered areas of the outcrop (photo **a**), where they look like big grains of salt on a pretzel. The quartz grains stand out because they resist weathering better than the other minerals do.

The dominant colour of the rock is pale grey-green, but some surfaces are stained red. This is due to weathering of iron in the rock.

Interrupting the colour scheme are a few bands of dark blue-green rock (photo **b**). They are slightly younger gabbro dykes. They formed as sheets of hot magma cut through the pre-existing tuff to feed lava flows higher in the volcanic pile (see site 9, Related Outcrops).

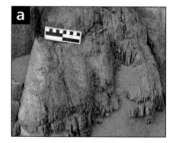

(**a**) Crystal tuff; (**b**) gabbro dyke.

1000   900   800   700   600   50

Z₁   Z₂   Z₃   €

FYI

- The ash deposits in Kennington Cove settled back to earth from a tall volcanic cloud or plume. This process is very different from the explosive, hot, pyroclastic flow deposits seen at some other sites. However, a single stratovolcano can make both kinds of deposit, and they are often interlayered with one another.

Ash and crystal tuff at Fourchu Head, Cape Breton Island.

Related Outcrops

Rocks like those at Kennington Cove make up most of the coastal outcrops in southeastern Cape Breton Island, including those at Fourchu Head. From Framboise, follow the Fleur de Lys Trail (St. Peters-Fourchu Road) to the community of Fourchu. Where the main road turns left, go straight (east) along the south side of the harbour. Park near the wharf (N45.71666 W60.24507), being careful not to obstruct wharf access. Continue on foot for about 600 metres. Near the breakwater, follow a track about 60 metres along the shore to outcrops (N45.71721 W60.23770) near a utility pole.

Pyroclastic blocks in tuff.

At this site, the pyroclastic rocks include ash and crystal tuffs similar to those at Kennington Cove. They also include some dramatically large pyroclastic blocks jettisoned from a nearby vent during a violent eruption.

Exploring Further

Kinley, Phosa. "Kennington Cove, Cape Breton." In *Impressions of Cape Breton*, edited by B. D. Tennyson. Cape Breton University Press, 1986.

Waves crash against the rocky shore in front of Louisbourg lighthouse. Salt spray keeps the well-exposed rock outcrops clear of lichen or plant growth.

Ending an Era
Waning Volcanic Activity in Avalonia

Some say that when Brigadier General Wolfe and his men took possession of this point in June 1758, it marked the beginning of the end of the siege of Louisbourg. The balance of power shifted, and change was in the air. Likewise, the rocks by the lighthouse signal the end of an era. They represent younger parts of the Coastal Belt, a final phase of explosive eruptions in this part of Avalonia.

On the ridge in front of the lighthouse and along the nearby Louisbourg Lighthouse Trail, you can view the remains of thick ash falls. Some outcrops include visible rock fragments that fell to earth after being ejected into the ash cloud. If you find unusually large fragments in an outcrop, it means the rock you are looking at formed relatively close to a volcanic vent.

By the time the very youngest rocks in the Coastal Belt formed (see Related Outcrops), the tall, violent volcanoes that covered this region with volcanic debris lay eroding and extinct, like the ruins of Louisbourg Fortress in the aftermath of battle.

Getting There

Driving Directions

Highway 22 becomes Main Street in the community of Louisbourg. From Main Street near the head of the harbour, watch for a turn (N45.92349 W59.96440) east onto Havenside Road. Follow Havenside Road for about 3.2 kilometres; it ends at the lighthouse.

Where to Park

Parking Location: N45.90674 W59.95747

Park in the gravel parking lot east of the lighthouse.

Walking Directions

Across the road from the lighthouse, a long rocky ridge rises along the shore. From the parking area footpaths lead onto, beside, and along the rock surface.

Notes

Although outcrops are accessible in normal tide conditions, high surf can create dangerous conditions along the shore.

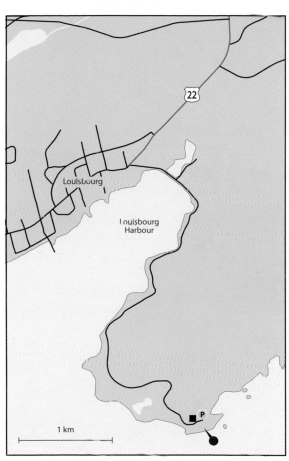

1:50,000 Map
Louisbourg 011G13

Provincial Scenic Route
Fleur de Lys Trail

On the Outcrop

The bare, abrasive surface of the shoreline ridge by Louisbourg lighthouse exposes layer upon layer of tuff, now tilted on edge.

Outcrop Location: N45.90616 W59.95734

As you walk from the parking lot across the rocky ridge toward the water, you are walking along the edges of layers that accumulated quickly, even catastrophically. As nearby volcanoes erupted violently, this pyroclastic debris sifted and splashed into the water of lakes that filled intervening valleys. Originally horizontal or nearly so, the layers were later steeply tilted by folding.

In some outcrops along the ridge, subtle bands of contrasting colour distinguish one pulse of ash from another. Many rock fragments within the ash matrix are rounded. Some of these (known as volcanic "bombs") erupted as blobs of liquid and cooled while travelling through the air (in contrast to angular fragments of pre-existing rock).

Volcanic bomb in ash.

Since the layers are tilted on edge, you may see examples where the ash layers drape around a fragment. The layers below were displaced when it fell, and those above later covered it.

1000	900	800	700	600	500
Z₁		Z₂		Z₃	€

FYI

- In Nova Scotia, Avalonian rocks of this age are restricted to Cape Breton Island. No volcanic rocks of equivalent age have been found in the Cobequid or Antigonish Highlands (see sites 2 and 3).

Related Outcrops

The Coastal Belt (shaded green in the map at right) is the youngest of three volcanic belts that make up this part of Avalonia (see sites 6, 7, and 8). The three belts are separated by major faults, so they probably formed in separate locations and were brought side by side later on.

Near Louisbourg, in Main-à-Dieu, boardwalks at the Coastal Discovery Centre (www.coastaldiscoverycentre.ca) lead to outcrops of a basalt flow (N46.00241 W59.84690). The basalt is a little younger than the volcanic rocks at Louisbourg lighthouse. Here, lava poured into valleys between dying volcanoes as tectonic plate motions changed direction, halting subduction.

An outcrop of basalt supports a lookoff platform along the boardwalk of Main-à-Dieu's Coastal Discovery Centre.

00   400   300   200   100   0

Є   O   S   D   C   P   Ŧ   J   K   Cz

The lookoff at Marble Mountain provides a view across West Bay of Bras d'Or Lake. The blocks of white marble at the lookoff are from the nearby quarry.

Ancient Margin

Marble Marking Ganderia's Proterozoic Margin

Look out across Bras d'Or Lake's West Bay from Marble Mountain to the far shore and picture this: Until about 400 million years ago an ancient sea sprawled here, separating two microcontinents—Avalonia, far away, and Ganderia, where you stand. The waters of West Bay conceal a major fault, a site of continental collision.

Early in its history, Ganderia was part of Gondwana and included a tectonically quiet continental margin edged by a warm, shallow sea. Sandstone and limestone formed, such as you might find in the Bahamas of today. Later tectonic events heated the rock, transforming it into quartzite and marble. Nothing like that can be found among the volcanic belts of Avalonia on the other side of the bay to the southeast. This sharp contrast in rock types led geologists to recognize the terrane boundary.

The Earth's tectonic engine is not the only force once active here. The community was an industrial powerhouse in days gone by. Hundreds of men worked to quarry the local marble and feed a gigantic mechanical crusher that rumbled and clanged through spring, summer, and fall each year.

Getting There

Driving Directions

From Highway 4 in Cleveland, northeast of Port Hawkesbury, follow County Line Road northeast to West Bay. In West Bay continue northeast on Marble Mountain Road, following it through Lime Hill to the community of Marble Mountain. The drive from Highway 4 in Cleveland to Marble Mountain is about 25 kilometres.

Where to Park

Parking Locations: (Lookoff) N45.82383 W61.03719
(Community centre) N45.81929 W61.04279
(Wharf) N45.82039 W61.03923

The community centre and lookoff are on Marble Mountain Road. For the public wharf, turn downhill on William MacInnis Lane, which intersects Marble Mountain Road less than 300 metres northeast of the community centre.

Notes

If you are travelling to the wharf, you may see signs restricting access to a private beach nearby. They do not pertain to wharf access.

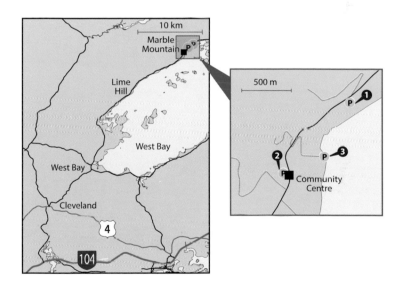

1:50,000 Map

Whycocomagh 011F14

Provincial Scenic Route

Bras d'Or Lakes Scenic Drive

On the Outcrop

A picnic area beside the Marble Mountain Community Centre provides an excellent view of the quarry face.

Outcrop Location: N45.81929 W61.04279

For a beautiful vista of Bras d'Or Lake as well as views of the community and its historic wharf, visit the lookoff along Marble Mountain Road. One reason you have a clear view from the lookoff is the absence of trees. This is due in part to the volumes of crushed marble still in place between the quarry and the shore. Also at the lookoff are large boulders of marble for close inspection.

The quarry is not accessible to the public, but from the Community Centre on Marble Mountain Road (waypoint and photo above) you can see the large quarry face. Boulders of

Marble boulder.

white marble from the quarry have been placed in the grounds of the centre.

Additional boulders, as well as a beach of white marble sand, are found at the wharf. The white colour of the marble is an indication of its purity. This made the rock particularly well suited for agricultural and industrial purposes.

1000	900	800	700		600	50
	Z_1		Z_2		Z_3	€

FYI

- Marble was quarried at Marble Mountain from 1869 to 1921. In the 1800s it was mainly converted to lime used in farming and for making mortar and plaster. In 1902, Dominion Steel and Coal Company bought the quarry and shipped crushed marble to Sydney for processing steel.

- The marble was originally limestone. Quartzite and slate found nearby were sandstone and shale. Together these rock layers formed in shallow water along the shore of the ancient continent of Gondwana.

Related Outcrops

Rocks from this early part of Ganderia's history are exposed along the central "spine" of Cape Breton Island. Metamorphosed sedimentary rocks (dark green), including the marble at this site, were intruded by igneous plutons (light green) as shown in the map below. Examples of the plutons can be seen at North River and Middle Head (site 11).

Kellys Mountain gneiss

The metamorphosed ancient sediments of Ganderia include the Kellys Mountain gneiss. It can be seen in a pull-off (N46.25189 W60.52882) on the north side of Highway 105 northeast of St. Anns. Despite a lot of graffiti, the folded dark and light layers are clearly visible in some boulders. The gneiss formed from a muddy sediment, and typically such rocks contain garnet. Instead, this gneiss contains cordierite, indicating that as the gneiss formed, it became hot without being buried very deeply. This may have happened as a rift in the crust allowed heat from the Earth's interior to rise into the crust.

Exploring Further

"This Was Marble Mountain," *Cape Breton's Magazine* 22 (1979), pp. 24–29 (available online at capebretonsmagazine.com).

500   400   300   200   100   0

Є   O   S   D   C   P   T   J   K   Cz

North River swirls and churns over outcrops of grey diorite near Little Falls in North River Provincial Park.

Basement Flooding

Tectonic Change on Ganderia's Margin

The footpath to Little Falls in North River Provincial Park leads down, down to the valley floor through the cool shadow of the woods. It's a bit like a trip down the basement stairs, and like a basement, the riverbed offers promise of discovery. Not Aunt Hattie's old butter churn, but a piece of history all the same.

About 560 million years ago, this part of Ganderia underwent a dramatic shift of tectonic plates. It had been a stable, quiet continental margin (see site 10), but now a subduction zone formed beneath it. Huge volumes of molten rock flooded upward from the subduction zone. The melt pooled in magma chambers that fed volcanoes at the surface.

The grey rock, diorite, that dominates the valley floor here preserves a record of those events. Rocks of similar age and character are found over a large area along the central axis of Cape Breton Island. They provide a glimpse of the foundation—known as the "basement"—underlying the many layers of sediment found in other parts of Ganderia, which stretches from New England to Newfoundland.

Getting There

Driving Directions

About 17 kilometres north of Highway 105 near St. Anns, follow Highway 30 (Cabot Trail) to North River Bridge. Turn (N46.30828 W60.62143) onto Oregon Road and continue for about 3.5 kilometres to the entrance for North River Provincial Park. Follow the park access road a further 400 metres to the parking area.

Where to Park

Parking Location: N46.31819 W60.66230

Park in the gravel area near the park kiosk.

Walking Directions

The trail head is on the left (south) as you enter the picnic and parking area. Follow the main trail downhill into the river valley and then for about 200 metres along the river until you reach the "Little Falls."

Notes

The provincial park is adjacent to the North River Wilderness Area, which has a more extensive trail system and includes North River Falls, Nova Scotia's tallest waterfall. See www.trails.gov.ns.ca/shareduse/vi017.html.

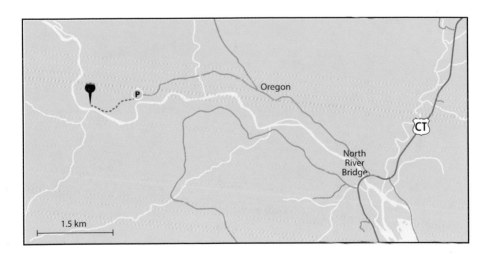

1:50,000 Map

St. Anns Harbour 011K07

Provincial Scenic Route

Cabot Trail

On the Outcrop

Grey diorite and younger pink granite are both scoured clean by the river, revealing plenty of detail.

Outcrop Location: N46.31681 W60.67059

Most rock outcrops near Little Falls are grey diorite. It contains a mixture of light minerals (quartz and feldspar) and dark minerals (biotite and amphibole), easy to see in the river-washed outcrops. Several features of the diorite developed as the molten rock cooled slowly.

Dark minerals are usually the first to crystallize. They sink through the remaining melt and gather in layers, which appear here as darker areas or bands. The rate at which molten rock cools affects the size of the mineral grains. Slow cooling creates larger grains and rapid cooling creates smaller ones. As the magma moves and circulates, it "stirs" the crystal mush, resulting in layers of contrasting coarse or fine texture.

Diorite. Note the subtle banding.

You'll also see outcrops and veins of pink granite and fine-grained black gabbro at this site. The granite is a younger rock related to the Cape Smokey granite (see site 12). The gabbro is even younger but of unknown age.

1000	900	800	700	600	500
	Z_1		Z_2	Z_3	Є

FYI

- The west coast of South America is a present-day example of a previously stable continental margin affected by tectonic plate subduction. The volcanic peaks in the Andes mountains formed as a result.

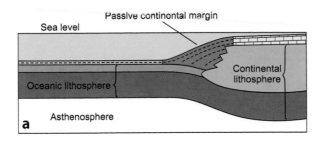

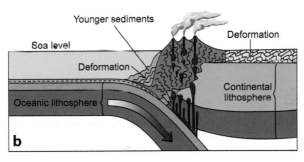

(a) Sediments (shown in purple) accumulate along a stable continental margin. (b) Subduction causes deformation and igneous activity in the older margin sediments.

- This type of tectonic setting creates igneous intrusions with a characteristic range of rock types, from diorite to granite. As a group, such intrusions have a chemical "signature," allowing geologists to identify their origin.

- An eye-catching feature at this site is a series of depressions in the rock, known as potholes. Often found in proximity to a waterfall, they form over decades or even centuries as rushing water rattles pebbles or boulders against the outcrop. In some, you may see a boulder that has been trapped and continues to spin and bump when the water is high, further deepening the hole.

A pothole near Little Falls.

Related Outcrop

On the headland at Middle Head, outcrops of diorite are plentiful. The diorite here contains numerous fragments, known as inclusions, of darker, finer-grained diorite (see detail).

Outcrop Location: N46.65469 W60.34994

Located by the Keltic Lodge near Ingonish, the trail on Middle Head is popular with hikers. At the eastern end of the trail, diorite can be seen in numerous grey outcrops. Like the diorite at North River Little Falls, it is part of a great influx of igneous intrusions forming the "basement" of Ganderia.

To reach the hiking trail, enter the Keltic Lodge grounds from the Cabot Trail and drive past the lodge to the parking lot at the trail head (N46.65542 W60.37222).

The first part of the trail leads across an area of pink rock, the Cape Smokey granite (site 12). Along the eastern end of the trail, grey diorite is more common. On the headland, the diorite contains numerous dark fragments. As at North River, they are evidence that dark minerals crystallize first as magma cools. Here they gathered into layers that were later disrupted by a new influx of molten diorite.

Diorite with inclusions.

1000   900   800   700   600   50

Z₁   Z₂   Z₃   €

By the Way

Near its midway point, Middle Head hiking trail passes through a grassy meadow.

Magma Mingling

Halfway along the Middle Head hiking trail is a sign showing how to circle back to the trail head or continue east to the headland. If you continue east about 200 metres, you'll find that the trail passes through a meadow (N46.65690 W60.36140) where the peninsula is especially narrow.

An unusual phenomenon is visible in cliffs below the north side of the meadow. Scan the grey diorite outcrops for a splash of intense pink (see photo). Between you and the obvious pink rock is a more complex-looking rock—part pink and part dark grey. It is a narrow intrusion, or dyke, in which two kinds of magma "mingled," flowing together without mixing. Like two fluids in a lava lamp, the pink granite magma and grey gabbro magma remained separate due to contrasts in their temperature and density.

With binoculars or a zoom lens you can see the tadpole-like blobs of gabbro within the granite—a feature characteristic of magma mingling. This complicated dyke cuts across and is much younger than both the diorite and the Cape Smokey granite on Middle Head.

Magma mingling along Middle Head trail.

500   400   300   200   100   0

€   O   S   D   C   P   Ʈ   J   K   Cz

Cape Smokey Provincial Park provides a wonderful panorama, including a pink cliff of the cape's Cambrian granite along the Cabot Trail.

Cambrian Landmark
A Familiar but Ambiguous Granite of Ganderia

Cape Smokey, so named because of the mist that often shrouds its summit, has been shaped by glaciers, streams, and ocean waves into an impressive landmark that is visible for miles around. Like many architectural landmarks of prominence and permanence, Cape Smokey is made of granite.

Until the 1930s, the Cape Breton Highlands were cut off to road transportation between Cape Smokey and Chéticamp. Earlier inhabitants found it easier to travel by water and used the cape's colourful cliffs as a navigational aid. Today's travellers follow the Cabot Trail, which winds its way past granite road cuts and climbs over this enduring landform. In the park at the top, you can inspect the rock first-hand.

The granite formed at the end of the Cambrian period, a time of rapid change as continents rearranged themselves. There are few igneous rocks of similar age in Ganderia, and they don't tell a consistent story. Some geologists think Ganderia began to break away from Gondwana around that time, causing igneous activity. Until more is known, the rock's geological past remains as smokey as the cape itself.

Getting There

Driving Directions

Along Highway 30 (Cabot Trail) about 13.5 kilometres south of Ingonish Beach, watch for signs to Cape Smokey Provincial Park. At the park access road, turn (N46.59532 W60.38555) east and follow it about 500 metres to the parking area.

Where to Park

Parking Location: N46.59313 W60.38073

The park road ends in a small loop by the gravel parking area.

Walking Directions

From the parking area, a number of footpaths lead through the brush to views of the coastline to the north and south. Outcrops and boulders can be seen along several of the paths.

Notes

The Cabot Trail climbs steeply from sea level to about 275 metres as you approach the park from either direction.

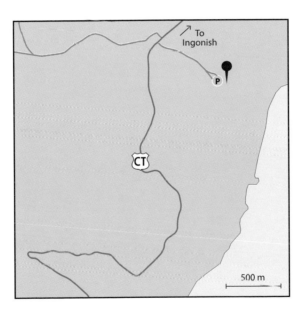

1:50,000 Map

Ingonish 011K09

Provincial Scenic Route

Cabot Trail

On the Outcrop

Outcrops and boulders of granite are easily found along the park's hiking trails.

Outcrop Location: N46.59294 W60.38008

Almost any trail in the park will bring you to a boulder or outcrop of Cape Smokey granite. To a geologist the term "granite" has specific meaning. A typical granite contains quartz and two kinds of feldspar in roughly equal proportions. It has less than 10 per cent of dark minerals such as biotite or hornblende.

Up close, quartz often appears light grey. One type of feldspar (plagioclase) is usually white or grey. The other (potassium feldspar) is usually pink. In certain angles of light, you can spot a feldspar crystal as the sun glints off a flat, shiny surface, since feldspar breaks in that characteristic way. Quartz breaks more like glass.

Cape Smokey granite.

The Cape Smokey granite has features that make it particularly durable. Dark minerals such as biotite weather easily, but this rock has almost none. The mineral grains are all similar in size, and they are randomly oriented rather than aligned, meaning there is no layering to weaken the rock.

1000   900   800   700   600   500

Z_1 Z_2 Z_3 €

FYI

- The age of the Cape Smokey granite was determined by crushing a sample and separating out tiny grains of zircon. The microscopic crystals are quite abundant in this rock. Uranium and lead in several individual grains were analyzed to pin down the age of 493 million years.

- The granite here is rich in silica but contains little iron and magnesium. That suggests the magma formed where continental crust was being pulled apart. In such a setting, heat flowing from the mantle melts continental material at great depths. This is happening below parts of New Zealand at present.

- The bare granite hills around Cape Smokey are the result of a forest fire in May 1968. With trees gone, the scant soil washed away, and as a result regrowth has been slow.

Magnificent scenery and picnic facilities invite visitors to stay awhile on Cape Smokey.

Related Outcrops

Cape Smokey is one of the few Cambrian intrusions (shaded green in the map at right) in this part of Ganderia. Outcrops of Cape Smokey granite are found along the Middle Head hiking trail near Keltic Lodge, and granite of similar age is found at North River (site 11). At Kellys Mountain a related granite intrudes the much older Kellys Mountain gneiss (see site 10).

	400	300	200	100	0
Є O	S D	C	P Tr	J	K Cz

A fault has brought steeply tilted layers of metamorphosed sedimentary rock into abrupt contact with the much younger black basalt of Pillar Rock.

Heat and Pressure
Metamorphism on the Margin of Ganderia

Where you exit the Cabot Trail for this site, the highway crosses a long, narrow lake. It separates the highlands from a sliver of rocky shoreline known as Presqu'île—"almost an island," as it is aptly named. The narrow valley occupied by the lake is bounded by two roughly parallel faults, one east and one west of the lake.

These faults extend under the beach and have caused a bit of geological mischief. The three rock types here—phyllite, sandstone, and basalt—have nothing to do with one another and are many millions of years different in age. Fault movements brought them together relatively late in the region's geological history.

The phyllite, which appears as great tilted slabs in cliffs below the highway, tells a long tale of tectonic drama. This metamorphic rock originated as muddy sediment, and later shale, along the margin of Ganderia. When Ganderia collided with Laurentia during the Silurian period, the shale was deeply buried, heated, and deformed to form phyllite. It was deformed again when Avalonia collided with Ganderia early in the Devonian period.

Getting There

Driving Directions

On Highway 30 (Cabot Trail) about 7.5 kilometres north of Petit Étang and just south of Presqu'île, watch for a beach that terminates in the black two-pointed landmark known as Pillar Rock. Turn (N46.68732 W60.95919) west into the entrance for the beach and follow a short gravel lane to the parking area.

Where to Park

Parking Location: N46.68656 W60.96002

Park in the open gravel area at the end of the access lane.

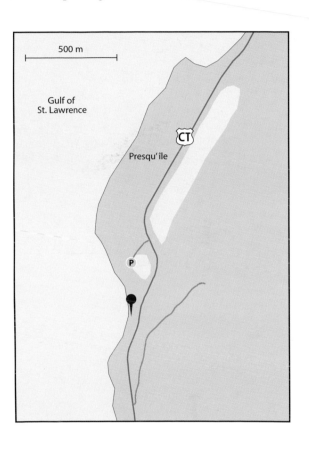

Walking Directions

From the parking area, make your way onto the beach and turn to the left, toward Pillar Rock. As conditions allow, follow the shore for about 200 metres until you reach the area below a stone wall reinforcement along the highway at the top of the cliff.

Notes

This site lies within the boundaries of Cape Breton Highlands National Park and requires a valid park pass.

1:50,000 Map	Provincial Scenic Route
Chéticamp River 011K10	Cabot Trail

On the Outcrop

Tilted nearly vertical and warped by folding, this rock known as phyllite breaks easily because its flat, flaky mineral grains are aligned.

Outcrop Location: N46.68517 W60.96012

Like the twisted trunk of an ancient tree, this rock looks like it has been through a lot, as indeed it has. The metamorphic fabric acquired long ago makes it susceptible to present-day weathering and wave action. Shiny broken slabs litter the beach beside the outcrop.

When shale is metamorphosed into phyllite, its clay minerals recrystallize into a type of white mica and similar minerals with flat, flaky grains. Like layers of fine pastry, the grains are aligned by tectonic pressures during metamorphism. Their alignment leaves the rock with a strong cleavage, that is, a tendency to split along parallel planes.

Phyllite cleavage surface.

Phyllite is similar to slate but has larger grains of mica, which make the rock distinctive. The abundant mica and related minerals in phyllite give its cleavage surfaces an easily scratched, soapy texture and a beautiful, silky sheen. Because of intense deformation, some cleavage surfaces are crinkled or wavy.

1000   900   800   700   600   500

Z_2 Z_3 €

FYI

- The rock here originated as mud that became shale and then phyllite. Present evidence suggests the mud was deposited on the margin of Ganderia while it was still attached to the supercontinent Gondwana late in the Ediacaran or early in the Cambrian period.

- Mineral characteristics indicate that the phyllite along the shore was buried to a depth of about 8 kilometres, reaching about 300°C. For schist and gneiss at the top of nearby French Mountain, the depth was 15 kilometres and the temperature 600°C. To expose such a great range of conditions over a relatively short distance, the crust must have been tilted—and perhaps also faulted—after the metamorphism occurred.

Related Outcrops

Metamorphosed sedimentary and volcanic rocks (shaded green in the map below) from this period of Ganderia's history cover a wide area of the Cape Breton Highlands—most of it very difficult to access. Examples are more easily found among beach boulders, for example, by Ingonish wharf (site 14).

Cap Rouge rest area.

You can visit other examples at the Cap Rouge rest area (N46.73095 W60.92405) at the base of French Mountain on the Cabot Trail. The stone wall around the lookoff contains several types of intensely metamorphosed rock from the highlands. Look for their shiny cleavage planes, some with bumps formed by distinctive metamorphic minerals such as garnet (red) and staurolite (brown).

By the Way

Like frozen waves about to crest, sandstone layers tilt gently toward the shore along the north end of the beach by Presqu'île.

Sandstone Cliffs

Along the north end of the same beach are low cliffs of tilted sandstone layers. Presqu'île itself is made of this rock type. The sandstone is of Carboniferous age and completely unmetamorphosed, so it has very little in common with the neighbouring phyllite or basalt.

Fault motions may have lowered the metamorphic rock of the highlands out of view here, leaving the overlying sandstone at surface level. Or fault motions may have slid this block of sandstone into position from the north or south.

The sandstone has some interesting features, including white veins of quartz and calcite (photo **a**). In some places, you may also find purple fluorite (photo **b**). Veins like these form when hot, mineral-rich water is flowing underground along cracks in the rocks. The minerals are deposited as the water cools.

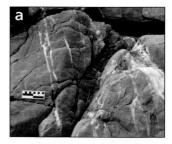

(**a**) Quartz and calcite veins; (**b**) fluorite in a vein.

By the Way

The sea stack known as Pillar Rock provides a distinctive landmark at the south end of the beach.

Basalt Sea Stack

A jagged sea stack of black basalt lends this site its name, Pillar Rock. A familiar type of landform, sea stacks are usually the result of differences in rock strength, with more resistant portions holding out against erosion.

This sea stack may owe its existence to the tectonic history of the area: Fault movements that brought the sandstone, basalt, and phyllite side by side may have left cracks in the basalt. Over time, such cracks would allow parts of the basalt to be loosened and removed by wave action.

More than just a picturesque landmark, the basalt hints at a restless period in the history of Ganderia as complex collisions involving Laurentia, Ganderia, and Avalonia reshaped the continents.

The basalt of Pillar Rock is the same age as granitic rocks seen at Green Cove headland (site 15). It was erupted along with the basalt at Grande Falaise (site 16) just a few kilometres away—another location where fault movements have altered the landscape.

Beach boulders near Ingonish wharf provide a rainbow of rock types for close inspection.

Ice-Age Hitchhikers
Metamorphic History as Told by Beach Boulders

You could thrash your way up stream valleys and through the underbrush all hither and yon over the Cape Breton Highlands … or you could visit the boulder-strewn beach near Ingonish wharf. Either way, you'll encounter many of the same rock types. During the last ice age, glaciers scoured the highlands and carried tons of rock out to the coast. Since then, waves have washed the rock clean, rounding and polishing them further as the boulders roll and clatter against one another.

The highlands of Cape Breton Island tell the story of the microcontinent Ganderia, where the early part of the Paleozoic era was an eventful time. Sedimentary and volcanic rock layers accumulated while Ganderia drifted toward Laurentia. Ganderia collided with Laurentia during the Silurian period. As a result, much of Ganderia was buried, deformed, and metamorphosed.

Rocks that crystallize at high temperatures become quite hard and strong—strong enough to survive glacial action and pounding surf. For that reason, the boulders found here are mainly medium- to high-grade metamorphic rocks (schist and gneiss) as well as resistant igneous rocks like granite and diorite.

Getting There

Driving Directions

In Ingonish, Highway 30 (Cabot Trail) runs east–west for about 3 kilometres along the shore of North Bay Ingonish. At the eastern end of this section of the highway, turn (N46.69168 W60.36176) onto Wharf Road and follow it to the public wharf and beach.

Where to Park

Parking Location: N46.68881 W60.35774

Park near the beach, being careful not to obstruct wharf traffic.

Walking Directions

Follow a short sandy track from the parking area to the beach.

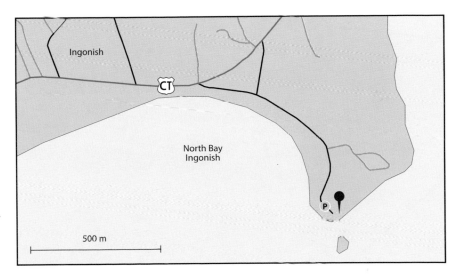

1:50,000 Map

Ingonish 011K09

Provincial Scenic Route

Cabot Trail

On the Outcrop

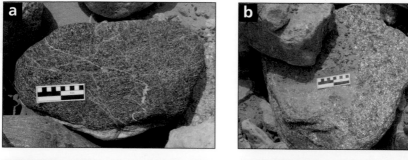

Boulders of (**a**) biotite schist, (**b**) muscovite schist, and (**c**) amphibolite are metamorphosed sedimentary and volcanic rocks originally deposited on top of older igneous rock such as (**d**) tonalite of the Ediacaran period.

Outcrop Location: N46.68863 W60.35725

Schist is any metamorphic rock with abundant mica or similar flaky minerals aligned by deformation. The alignment creates a layered, "platy" fabric. Typically the individual mica grains in schist are large enough to see. Light glints off their mirror-like surfaces, so look for boulders that have a sparkly appearance. The mica schist here is mostly rich in either black biotite or colourless muscovite (photos **a** and **b**).

Gneiss typically has a banded appearance. A common type of gneiss, amphibolite, is dark grey since it forms from dark igneous rock like basalt or gabbro. Some amphibolite boulders have eye-catching white bands of quartz and feldspar (photo **c**). During deformation when the rock was hot and pliable, a quartz-feldspar melt was injected into the darker amphibolite, forming the bands.

Another type of gneiss forms when a light-coloured, quartz- and feldspar-rich igneous rock such as granite or tonalite (photo **d**) is heated and deformed. Any flat, flaky minerals in the rock align when they recrystallize, creating subtle, speckled layering. Sometimes light and dark minerals gather into bands of contrasting colour.

1000   900   800   700   600   500

Z_1 Z_2 Z_3 ε

FYI

- As shown by field studies and laboratory experiments, a mud-rich sedimentary rock, shale, buried deeper and deeper in the Earth will undergo a series of changes resulting in different rock types.

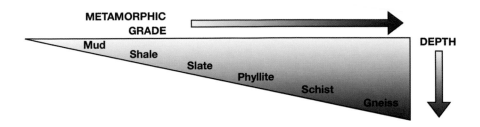

- Much of the rock you now see as schist and gneiss by Ingonish wharf probably had its origin in a chain of volcanic islands. During the Ordovician or Silurian period such a setting could have developed right beside Ganderia as it approached Laurentia. When Ganderia and Laurentia collided, the volcanic and sedimentary rock of the island chain was caught up in the collision, deeply buried, and metamorphosed.

Related Outcrops

Phyllite along the shore at Pillar Rock (site 13) was metamorphosed during the same event, but at lower temperatures and pressures compared to the beach boulders near Ingonish wharf. Note that many of the pink boulders used to make the breakwater near the wharf are not from here, but are related to granite outcrops seen along the shore between Ingonish and White Point (see site 15).

The outcrops of layered, grey rock on the shore at Ingonish wharf are limestone from the Carboniferous period. The limestone formed in the ancient Windsor Sea and is related to limestone, gypsum, and salt deposits elsewhere in Nova Scotia (see site 37).

Exploring Further

For more information about the movements of ice-age glaciers in Nova Scotia, see the Nova Scotia Department of Natural Resources' *Reading Room 1: The Story of Glaciers in Maritime Canada* (www.novascotia.ca/natr/meb/fieldtrip/glacier.asp).

500   400   300   200   100   0

€   O   S   D   C   P   Ŧ   J   K   Cz

Formerly the site of a fishing community, Green Cove is sheltered by headlands of pink Devonian granite.

Crustal Melting
Granite Resulting from Continental Collision

Forest, bird song, rock, sea, and sky bring most visitors to the pristine beauty of Green Cove. Nature has had its way; no obvious trace remains of those who lived here more than 100 years ago. In the late 1800s nearby towns became crowded during a boom in the fishing industry. Even the rocky land along little Green Cove was settled in the quest for lobster and cod.

Those hard-working fishermen and their families may have wished for better soil in the garden, but they probably had no time to wonder further about what lay underfoot. It took the work of later, scientific pioneers to discover that the granite here has its own, more ancient story to tell.

The Devonian period was a time of tectonic chaos for the microcontinents of Ganderia and Avalonia. Plate motion drove them grinding and jostling against one another over millions of years. Portions of continental crust were forced to great depths and some melted, yielding granite magma. This extensive, well-exposed outcrop provides professionals and amateurs alike with a textbook-quality example of these granite intrusions.

Getting There

Driving Directions

Along Highway 30 (Cabot Trail) about 8.5 kilometres from the sharp turn at Neils Harbour, watch for signs for Green Cove. Turn (N46.75084 W60.32543) east onto the access road and follow it about 50 metres to the parking area. There are several related sites nearby; see Exploring Further for details.

Where to Park

Parking Location: N46.75108 W60.32488

Park at the north end of the parking area for the most direct access to the trails.

Walking Directions

From the parking area follow the trail, which includes sections of boardwalk and stairs, through the trees and onto the rocky headland.

Notes

This site lies within the boundaries of Cape Breton Highlands National Park and requires a valid park pass.

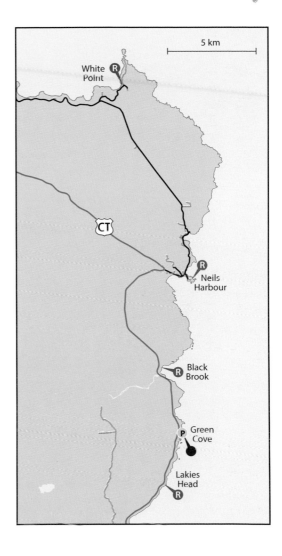

1:50,000 Map
Dingwall 011K16

Provincial Scenic Route
Cabot Trail

On the Outcrop

Like an ancient plaid of pink and grey granite, outcrops of this complex intrusion are criss-crossed by dykes of several kinds.

Outcrop Location: N46.75064 W60.32407

The wave-washed surface of the outcrop reveals many details of colour and texture. The intrusion is a complex one and it's easy to identify several different varieties. The most common granite contains small flakes of a dark mica, biotite.

Cutting across the biotite granite are irregular dykes and patches of aplite, a pale pink rock containing quartz and feldspar but little or no biotite. It has a fine, sugary texture. In contrast, pegmatite dykes are darker pink with large, clearly visible crystals of quartz, feldspar, and mica (biotite and muscovite).

Areas of biotite schist are likely fragments of older crust carried along in the intrusion as it moved upward. Fragments of another older rock type, the Cameron Brook granite, are easily recognized by their large, well-defined crystals of pink feldspar. This distinctive rock type was also caught up in the biotite granite magma as it moved through the older crust.

Cameron Brook granite.

1000   900   800   700   600   500

Z₁   Z₂   Z₃   €

FYI

- The very large crystals in pegmatite dykes form as water in the magma enables crystals to grow more quickly. Aplite dykes likely represent dry areas where water had previously escaped from the magma.

- Large crystals in pegmatite have a variety of industrial uses. In the past, for example, huge sheets of transparent mica were mined near Moscow, Russia, for use as window panes. The term "Muscovy glass" led to that mineral's formal name, muscovite.

- Based on uranium and lead analyses of their zircon mineral grains, the biotite granite as well as related aplite and pegmatite at Green Cove are all 375 million years old. Granite plutons in this age range are found not only in Ganderia and Avalonia but also in Meguma (see sites 29 and 30).

- Some of these Devonian granites, including the granite here at Green Cove, contain both biotite and muscovite. This and other characteristics show that they formed by melting of metasedimentary rocks in the deep crust.

A boardwalk leads to shoreline outcrops of Devonian granite at Lakies Head.

Related Outcrops

A large volume of granite and granitic gneiss (shaded green in the map at right) is exposed along this section of coastline. Similar rocks appear all the way from Lakies Head (N46.73504 W60.33199) to Black Brook Provincial Park (N46.77602 W60.33336), Neils Harbour (N46.80677 W60.31813), and White Point (trail head, N46.87644 W60.35228), as well as in boulders by Ingonish wharf (site 14).

500   400   300   200   100   0

€   O   S   D   C   P   T   J   K   Cz

At the Grande Falaise picnic park near Petit Étang, a fault lies between dark basalt, below, and pinkish granite, above.

Tectonic Pileup
Aftermath of Plate Collision in Ganderia

After Ganderia collided with Laurentia during the Silurian period, it was stuck, as they say, between a rock and a hard place. Coming along behind it were the microcontinents of Avalonia and then Meguma, in turn squeezing Ganderia against Laurentia as the continent grew.

At Grande Falaise, you can see evidence of the tremendous forces at work during this time of continental rearrangement. The "great cliff," as its name translates, stretches for 600 metres beside the Cabot Trail north of Petit Étang. Visible in places at the base of the cliff is a dark layer of basalt that was erupted onto the surface of the Earth about 375 million years ago—around the time Avalonia collided with Ganderia. As the basalt cooled, there was nothing above it but sky.

Small problem: Above the basalt today there sits an immense mass of pinkish-grey granite. It originally formed far below and intruded into the crust 550 million years ago. A thrust fault between the two rock types explains this pileup. The older granite was pushed upward and westward over the basalt, probably early in the Carboniferous period.

Getting There

Driving Directions

On Highway 30 (Cabot Trail) between Presqu'île and Petit Étang, as you come in view of a prominent rocky cliff on the east side of the highway, watch for signs for the Grande Falaise picnic area. Turn into the access lane.

Where to Park

Parking Location: N46.67430 W60.95330

Park at the end of the lane in the open gravel area.

Walking Directions

You don't really need to leave the grassy area for excellent views of the cliff face. However, a trail on the north side of the little park provides closer access.

Notes

Rock debris at the base of the cliff is unstable. This site lies within the boundaries of Cape Breton Highlands National Park and requires a valid park pass.

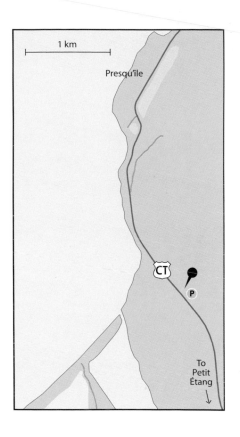

1:50,000 Map

Chéticamp River 011K10

Provincial Scenic Route

Cabot Trail

On the Outcrop

Narrow, dark grey dykes (by arrows) slice at a low angle across the granite near the top of the cliff. In this view, a thick pile of broken granite fragments conceals the basalt underneath.

Outcrop Location: N46.67453 W60.95338

The granite in the upper part of the cliff is shattered by numerous cracks and fractures, probably due to its trip along the thrust fault. Over time a huge wedge-shaped pile of broken pieces has accumulated along the base of the cliff, in most places obscuring the basalt below it. This loose material is known as scree or talus. It is not stable and is unsafe to climb on.

Even from ground level, you can see narrow slashes of dark grey crossing the exposed upper portions of the cliff. These are dykes related to the younger basalt now located below. Magma flowed through older rock formations, including the granite, on its way to the surface—when the basalt was still above it.

In some parts of the cliff face you can see a narrow zone of brick red sedimentary layers between the basalt and the granite. They accumulated on top of the basalt prior to arrival of the granite. Veins of white calcite appear near the base of the granite. These minerals were deposited by fluids moving along the weakened fault zone.

1000	900	800	700	600	50
Z_1		Z_2		Z_3	$\in$

FYI

- The basalt at this site and the granite at Green Cove (site 15) were both associated with the complex collision of Avalonia and Ganderia. In some places where collision caused the crust to thicken, its lower regions melted forming granite magma. In other places at the same time, the crust was being stretched, thinned, and torn, allowing basalt magmas to rise from the mantle.

- Thrust faults form when the Earth's crust is being compressed. In (a), the basalt (green) is located at the land surface shortly after its eruption. The older granite (pink) is underneath it. In (b), movement on a thrust fault has pushed the granite sideways and upward, over part of the basalt. The dotted line shows how erosion might create a new land surface, exposing the granite, fault, and basalt as they are seen today at Grande Falaise.

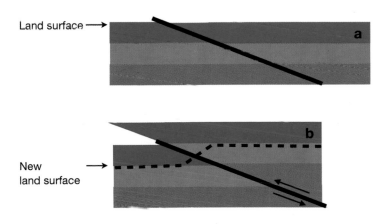

Related Outcrops

Volcanic eruptions (shaded green in the map at right) related to the basalt at Grand Falaise occurred in a number of locations in western Cape Breton Island during the Devonian period. The basalt in the sea stack called Pillar Rock (site 13) was also part of this event.

The granite at Grande Falaise is about the same age as the diorite at North River and Middle Head (site 11). They all formed late in the Ediacaran period.

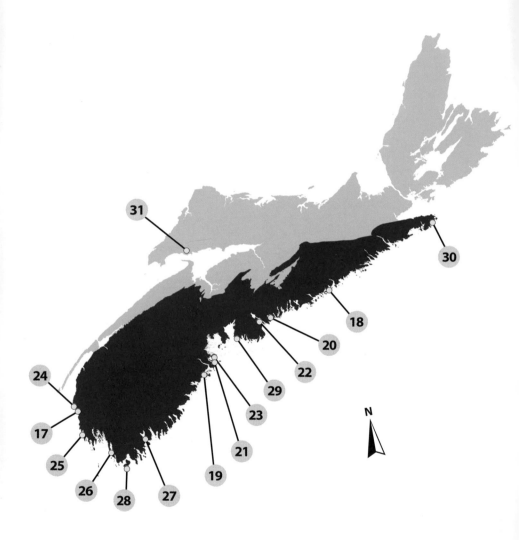

MEGUMA

Locations

Mainland Nova Scotia south of the
Cobequid-Chedabucto fault system

Origin

Continental margin of Gondwana

Key Features

Thick layers of sea-floor sediment
High-temperature metamorphism
Gold and other minerals
Large volumes of granite

At these sites, you can ...

17	**Bartletts Beach**	Encounter Meguma's oldest known rock layers.
18	**Taylor Head**	Witness the results of underwater landslides.
19	**Green Bay**	Learn how a rock can look so much like driftwood.
20	**Rainbow Haven**	Track down a marker horizon vital for geologic maps.
21	**The Ovens**	Catch gold fever at an historic mining location.
22	**Point Pleasant**	Examine sand ripples from an ancient ocean floor.
23	**Blue Rocks**	Critique the artistry found in the hinge of a fold.
24	**Cape St. Marys**	Trace an unconformity marking change in Meguma.
25	**Cape Forchu**	Venture through the remains of volcanic eruptions.
26	**Pubnico Point**	Appreciate the versatility of metamorphosed mud.
27	**Sandy Point**	Take Earth's temperature using metamorphic minerals.
28	**The Hawk**	Admire the beauty of a rock so hot it started to melt.
29	**Peggys Cove**	Enjoy a famous view while studying Meguma granite.
30	**Black Duck Cove**	Stroll by nature's quarry, an exfoliating granite.
31	**Wharton**	Visit a fault scarp related to Meguma's collision.

The Meguma terrane is unique to Nova Scotia—nowhere else can it be seen on land. Offshore it extends across the continental shelf from southern New England past Nova Scotia toward the Grand Banks.

Meguma's story began late in the Ediacaran period, when the sea flooded a rift valley on the margin of the supercontinent Gondwana. The rift's steep sides eroded rapidly and sediment was carried onto the sea floor by powerful turbidity currents, akin to underwater landslides.

Bartletts Beach (site 17).

The Goldenville Group of turbidite layers records the earliest known history of Meguma beginning about 540 million years ago. Though chaotic, turbidity currents leave predictable and often beautiful signs of their passage (sites 17, 18, 19). About 500 million years ago, changes in underwater conditions caused a distinctive manganese-rich layer to form. It now marks the top of the Goldenville Group (site 20), so named because many of Meguma's turbidites host deposits of gold (site 21).

Over time, sediment had to travel farther to reach Meguma as sea level rose and erosion flattened the adjacent land. In consequence, Meguma's younger Halifax Group of turbidite layers is fine-grained—rich in silt and mud but poor in sand. Oxygen-starved sea-floor conditions (site 21) ended when the Rheic Ocean began to open about 485 million years ago (sites 22, 23).

Green Bay (site 19).

Cape Forchu (site 25).

Following a long gap in Meguma's geologic record, rocks above the turbidites reflect a very different environment—a shoreline setting of shallow water near land. A thick pile of lava and volcanic ash buried parts of Meguma (sites 24, 25) about 440 million years ago as the Rheic Ocean narrowed and the terrane drifted northward toward Laurentia.

About 390 million years ago, all Meguma's layers of sedimentary and volcanic rock were folded and metamorphosed during the early stages of Meguma's collision with Avalonia. In a few areas, high-temperature minerals formed (sites 26, 27) and some rocks partially melted (site 28).

Probably as a result of subduction beneath Meguma during closure of the Rheic Ocean, huge volumes of granite magma formed in the deep crust, mainly between 380 and 372 million years ago (sites 29, 30). Apparently the terrane was no longer being deformed by that time: The granite plutons bear little sign of the folding and cleavage so obvious in Meguma's older rocks.

Meguma's journey as a lone microcontinent ended as it moved beside Avalonia along a transcurrent fault, probably between 360 and 340 million years ago. A complicated fault zone remains to mark the boundary between them (site 31).

Peggys Cove (site 29).

Low outcrops of sandstone are an inconspicuous but significant feature of the 2.5-kilometre sweep of Bartletts Beach north of Port Maitland.

Rift Valley
Oldest Known Outcrop of Meguma

Where is Nova Scotia's oldest tree? How old was the oldest lobster ever caught off Nova Scotia's shores?* A fascination with record-setting age extends to the world of geology, too. On Bartletts Beach you can visit the oldest known rock from an ancient microcontinent, Meguma. Remnants of Meguma underlie much of mainland Nova Scotia.

When these oldest rocks of Meguma formed, the region lay at the bottom of a deep sea caused by rifting along the edge of Gondwana. The rift was tectonically active, so it had high, steep sides. Huge volumes of rapidly eroded sediment were carried into the sea and piled precariously onto steep underwater slopes.

Periodically an underwater landslide would send many cubic kilometres of sediment rushing downslope in what is known as a turbidity current. Time after time, turbidity currents swept down the sides of the narrow sea, spreading their massive loads of sediment in layers on the sea bottom.

---

* A 425-year-old tree has been found in an old-growth forest in southwestern Nova Scotia. The largest lobster caught to date was about 100 years old.

Getting There

Driving Directions

From Highway 1 about 20 kilometres north of Yarmouth (between Port Maitland and Salmon River), watch for Bartlett Shore Road on the west side of the highway. Turn (N44.01494 W66.14941) west and follow the road for about 650 metres. It ends near the beach.

Where to Park

Parking Location: N44.01619 W66.15785

Park in the open area at the end of Bartlett Shore Road.

Walking Directions

From the parking area, walk down onto the beach. As you face the water, the outcrops are to the left (south). Walk about 200 metres along the shore until you reach a series of small low outcrops in the sand.

Notes

The outcrops are best accessed during low tide conditions. Because storms shift the distribution of sand, the location and extent of the exposed rock may vary. The beach that extends north from the parking area is designated a Protected Beach under provincial law.

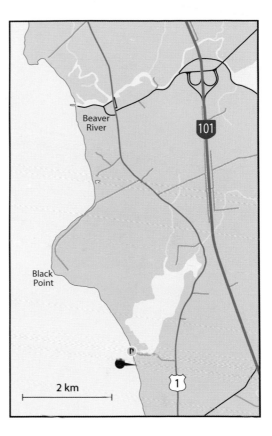

1:50,000 Map

Meteghan 021B01

Provincial Scenic Route

Evangeline Trail

On the Outcrop

Storms and currents shift the sand of Bartletts Beach from time to time, so the exposures you see may differ from those shown here.

Outcrop Location: N44.01442 W66.15752

The rock here is a type of metamorphosed sandstone. The individual sand grains include quartz, feldspar, and tiny rock fragments. Between the grains is a very fine-grained matrix that was originally mud but now includes mica and other flaky minerals.

On some outcrops you will find subtle, nearly horizontal bands of darker and lighter grey (see detail). These bands formed as successive swirls and clouds of the sediment-laden current dropped new layers of sand. Long afterward, the layers along this part of the coast were folded into an arch-like structure known as an anticline. Because this site is located at the crest of the fold, the original sedimentary layering is still approximately horizontal (see FYI).

Sedimentary layers cut by later near-vertical cleavage.

Cutting across the sedimentary layers at a high angle are cracks that break the rock crudely into plate-like fragments. This cleavage, as it is known, formed during the later folding and metamorphism.

1000	900	800	700	600	500
Z_1		Z_2		Z_3	€

FYI

- In all, Meguma's turbidites total more than 12 kilometres in exposed thickness. At right, the arrow indicates this site's position in the sequence.

- The rocks both north and south of this site are younger due to folding of the rock layers in an anticline. In the diagram at left below, (**a**) shows the sedimentary layers deposited in sequence, with the first, oldest layer at the bottom. In (**b**), notice how folding and later erosion expose the oldest layer at the centre of the anticline.

- In (**b**) below, also notice how at the very top, or crest, of the anticline, each rock layer remains horizontal (like the ones on Bartletts Beach), while along the sides of the anticline the layers are tilted.

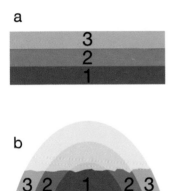

Related Outcrops

The turbidites of the Goldenville Group (shaded dark blue in the map above right) are widespread in southern and southwestern regions of the mainland.

At Taylor Head and Green Bay along the south shore (sites 18, 19), you can see other sedimentary rock features associated with turbidity currents in the same rift-related sea.

500	400	300	200	100	0
Є	O	S	D	C	P	R	J	K	Cz

This low, rounded outcrop of metamorphosed sandstone in Taylor Head Provincial Park was shaped by glaciers during the last ice age.

Bouma Sequence

Graded Bedding Created by Turbidity Currents

During the Cambrian period, sediment many kilometres thick accumulated in Meguma's sea. For 50 million years, turbidity currents—dense, gravity-driven clouds of water, sand, silt, and mud—rushed down steep underwater slopes and spread over the sea bottom.

Though catastrophic in origin and seemingly chaotic in action, turbidity currents have very predictable outcomes. In a stroke of insight in 1962, sedimentologist Arnold Bouma described a distinctive sequence of sedimentary rock layers common to many deepwater deposits. The relative thickness of the layers may vary, depending on the sediment involved, but their order is predictable.

The sequence now bears his name and allows turbidites, as these rocks are known, to be readily recognized. A Bouma sequence begins with a thick layer, or bed, of graded sandstone. In graded bedding, sand grains are sorted by size from coarser base to finer top. Another part of the Bouma sequence is a layer of siltstone, often with complex, swirling or rippled laminations. At this site, you can see examples of both these rock types, so typical of Meguma's long history of deepwater sedimentation.

Getting There

Driving Directions

On Highway 7 about 11 kilometres southwest of Sheet Harbour and just east of Spry Bay, watch for signs for Taylor Head Provincial Park. At the park entrance, turn (N44.84457 W62.58163) south onto Taylor Park Road. Follow the road for almost 5 kilometres to the parking area.

Where to Park

Parking Location: N44.80745 W62.56190

Park in the gravel parking area at the end of Taylor Park Road.

Walking Directions

A gravel and boardwalk trail leads from the eastern side of the parking area about 150 metres to the beach. In the gazebo along the way you may find a logbook that visitors can sign. On the beach turn left (north) and walk about 400 metres to a series of low outcrops in the sand.

Notes

The outcrop is best visited in low tide conditions but may be accessible at other times. The walking distance is from the parking area to the northern outcrop and back. Similar outcrops occur at the other end of the beach.

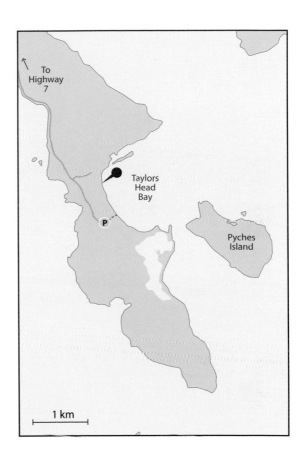

1:50,000 Map

Tangier 011D15

Provincial Scenic Route

Marine Drive

On the Outcrop

A pattern of alternating layers is characteristic of turbidite deposits like this one. Later folding and metamorphism created the cleavage that cuts vertically across them.

Outcrop Location: N44.81153 W62.56212

The most interesting feature of this sandstone is best seen at the southern end of the outcrop. On a vertical rock face, sedimentary layers appear as alternating bands of lighter and darker grey. Each pair of beds represents a single (though incomplete) Bouma sequence—the work of a single turbidity current.

The lower, dark grey bed in each pair is sandstone; the upper, light grey bed is finer-grained siltstone. In some places you can also see traces of very fine-grained blue-grey shale above the siltstone. Then, abruptly, a new layer of coarse sandstone begins, representing the work of the next turbidity current.

The sandstone beds are graded, as your sense of touch can tell you. The rock surface feels quite gritty at the base of the bed and less so at the top, like different kinds of sandpaper. Grading occurred as the turbidity current gradually lost energy and slowed down. The largest grains, being heaviest, were the first to stop moving. Smaller grains remained suspended longer, settling more slowly.

FYI

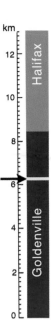

- In all, Meguma's turbidites total more than 12 kilometres in exposed thickness. At right, the arrow indicates this site's approximate position in the sequence.

- On the top surface of this outcrop, in certain angles of light, you can see subtle, parallel grooves (striations). They were scratched as advancing glacial ice scraped rock fragments across the outcrop during the last ice age about 20,000 years ago.

- A complete Bouma sequence consists of five parts. Each part represents a phase in the history of a single turbidity current as it travels downslope, losing energy in the process. The rocks (turbidites) left behind to record such events often don't preserve the whole sequence, but the parts that are preserved always occur in the same order.

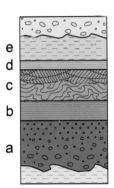

Sketch showing the five parts of an idealized Bouma sequence:
(a) rapidly deposited, graded sand, sometimes with a gravel base;
(b) thin, parallel laminations of sand;
(c) rippled or convoluted layers of silt;
(d) thin, parallel laminations of silt;
and (e) mud.

Related Outcrops

To see other turbidite features in the park, follow Spry Bay Trail to the observation tower on the west side of Taylor Head. The stairs to the tower follow the top surface of a sandstone layer. It has a messy, bumpy appearance because it is covered with small mounds known as sand volcanoes. They formed as the layer was quickly buried by more sediment. Excess water was pressed out and escaped through tube-like vents, erupting sand with it.

Looking north from the viewing deck, you can see layer upon layer of sandstone along the shore, all part of the immensely thick Goldenville Group. For other examples of Meguma's deepwater sandstones, see sites 17 and 19.

Exploring Further

For a video of a turbidity current produced in an oceanographic laboratory, visit activetectonics.coas.oregonstate.edu/model_turbs.htm.

500   400   300   200   100   0

€   O   S   D   C   P   T   J   K   Cz

On the beach at Green Bay near Petite Rivière, narrow fingers of bedrock extend into the water. Note large boulders placed to control erosion were brought from another location.

Swirling Sand
Story of a Changing Landscape

No major upheavals affected Meguma during the Cambrian period. For 50 million years a sea covered it; for 50 million years sediment poured into the sea and fell in tumbling currents across its floor. However, based on recent advances in mapping (see site 20) and other discoveries, geologists have recognized subtle changes in Meguma's environment over time.

The sea that covered Meguma began as a steep-walled rift on the margin of Gondwana. Initially, high-energy turbidity currents swept across the sea floor (see site 17) carrying sediment from the adjacent rift. But by the time the sedimentary layers at Green Bay were deposited, the rift was no longer active. The walls of the rift became less steep. Rivers feeding sediment into the sea began reaching into Gondwana's interior.

At Green Bay, beautifully sculpted outcrops that look more like driftwood than rock were deposited by turbidity currents travelling with medium-energy flow (see site 18, FYI). Layers like these became thicker and more common in Meguma around the middle of the Cambrian period, one sign that the environment was slowly changing.

Getting There

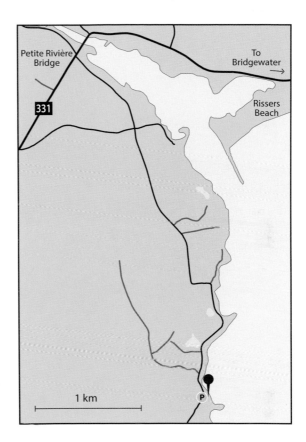

Driving Directions

Highway 331 south of Bridgewater passes through Rissers Beach Provincial Park and then turns sharply southwest to cross Petite Rivière. Less than 100 metres from the river, in the community of Petite Rivière Bridge, turn (N44.23333 W64.44725) southeast onto Green Bay Road. Follow the road about 2.7 kilometres. You will see a line of large boulders standing between the road and a small beach.

Where to Park

Parking Location: N44.21292 W64.43665

Park in the open area by a small canteen, being considerate of others' need for access.

Walking Directions

From the parking location, walk between boulders onto the sandy beach. The outcrops are near, but slightly left (north) of, the access point.

Notes

The beach that extends south from this location is designated a Protected Beach under provincial law.

Petite Rivière Bridge

To Bridgewater →

331

Rissers Beach

1 km

P

1:50,000 Map

LaHave Islands 021A01

Provincial Scenic Route

Lighthouse Route

On the Outcrop

(a) Different weathering rates highlight the layering in a series of turbidite deposits. (b) Curved laminations in fine-grained sandstone originated as sand ripples formed by a medium-energy turbidity current.

Outcrop Location: N44.21294 W64.43637

Look for the long, narrow outcrop that stretches from the boulder retaining wall toward the water (photo **a**). The fine sand, silt, and mud that form these rock layers were deposited by a series of turbidity currents. Later metamorphosed and folded, they are now on edge. The top of each layer is toward the left as you face the water.

The outcrop is deeply grooved because the sandstone layers are more resistant to weathering than the others. Weathering on the surface of the sandstone has revealed many fine, curved mini-layers known as cross-laminations (photo **b**). Because of this, some parts of the outcrop look more like driftwood than rock. Cross-laminations form as a current flows across a bed of fine sand, creating ripples. Wind creates sand dunes in the desert in a similar way, but on a much larger scale.

The finer-grained layers here lack cross-laminations. That's because the tiny particles of silt and mud were more easily suspended in the water and settled hours or days later, after the turbidity current had passed by.

1000   900   800   700   600   5

Z₁   Z₂   Z₃   €

FYI

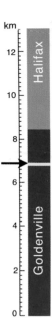

- In all, Meguma's turbidites total more than 12 kilometres in exposed thickness. At right, the arrow indicates this site's approximate position in the sequence.

- Tiny zircon grains eroded from rocks can resist destruction and travel with other sand particles on their way to the sea. Zircon grains from ancient areas of Gondwana—2,000 to 3,000 million years old—are found in Meguma sediments of this age and younger but not in older layers. This suggests that the rivers feeding Meguma's sea were reaching farther into Gondwana's interior.

- Cross-laminations form as water currents pass over a rippled, sandy surface. Sand grains roll or bounce up the back of a ripple, but fall down the far side and accumulate there. Over time, a ripple moves in the direction of the current as laminations build up along its steep downstream face. If the sand is coarser and the layers wider, this feature is known as cross-bedding instead.

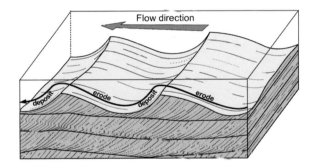

Related Outcrops

Sediment deposited in the sea covering Meguma became finer grained over time. You can see older, thick beds of relatively coarse sandstone at Bartletts Beach (site 17). Alternating beds of sandstone and siltstone at Taylor Head Provincial Park (site 18) are slightly older than the rocks here. Younger examples at Rainbow Haven (site 20) continue the trend with more beds being relatively fine grained.

Exploring Further

For an animation of cross-laminations being formed in ripples of sand, visit the U.S. Geological Survey's website at walrus.wr.usgs.gov/seds/bedforms/animation.html.

Rainbow Haven beach arcs around the west side of the bay, providing a fine view back toward the provincial park.

Dividing Line
A Unique Marker Horizon for Meguma

Geologists mapping Meguma in earlier centuries were impressed by the unrelenting sameness of the turbidite layers. They encountered sandstone, siltstone, and shale in various combinations over and over from Canso to Yarmouth. Not sure how to recognize subtle changes (see site 19), they wore down their coloured pencils shading large tracts a monotonous hue.

Just picture trying to make a geological map of these rather uniform layers. You find an outcrop along a beach … You follow it, but sand and trees hide the rest … The next beach provides another glimpse … Rocks in a stream bed give you just a peek. You gather pieces of the puzzle but you're not sure how they fit together, especially because the rock layers have been folded into complex shapes.

You need what geologists call a marker horizon: a distinctive, thin, widespread layer you can always recognize. Here beside Rainbow Haven beach, you can see the marker horizon used by Nova Scotia geologists to create groundbreaking maps of Meguma's thick sedimentary pile. New, precise maps have led to significant insights about Meguma and the sea that covered it.

Getting There

Driving Directions

Follow Highway 207 (Cole Harbour Road) in the community of Cole Harbour and watch for Bissett Road. Turn (N44.67250 W63.47800) southeast onto Bissett Road and travel about 5.5 kilometres—it ends at Cow Bay Road. Turn left (north) onto Cow Bay Road, but after about 100 metres turn right (east) into Rainbow Haven Beach Provincial Park.

Where to Park

Parking Location: N44.64782 W63.42274

Park in the gravel lot on the left near the park entrance.

Walking Directions

From the parking area, continue a short distance along the park access road then bear right onto a wide gravel footpath. Follow the path and then a boardwalk onto the beach, then turn right (south) and walk along the sand and cobble beach, following the shoreline for about 500 metres.

Notes

Beyond the first rock outcrop, this walk will take you outside the boundaries of the provincial park. Please stay on the shore and respect the privacy and property of adjacent landowners.

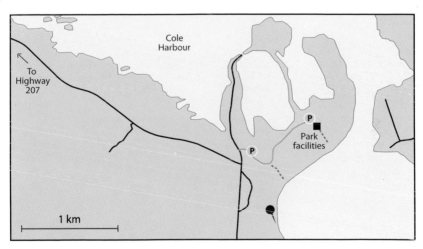

1:50,000 Map

West Chezzetcook 011D11

Provincial Scenic Route

Halifax Metro

On the Outcrop

Vertical cleavage cuts across the near-horizontal sedimentary layering. The thin, black layers are rich in manganese.

Outcrop Location: N44.64198 W63.42003

The first few outcrops you encounter along the beach are smooth, rounded sandstone. Farther along you'll see jagged-looking grey and brown slate, much of it broken into upright slabs by near-vertical cleavage (formed during much later folding). Bands of contrasting colour or texture extend across these outcrops. The bands are the original sedimentary layers.

The marker horizon is a series of sedimentary layers rich in manganese. In some outcrops, you may see what look like strings and knots of black running parallel to the sedimentary layering, as seen in the photo above. These are concentrations of manganese oxide. Iron occurring with the manganese produces an orange-brown stain on some outcrops.

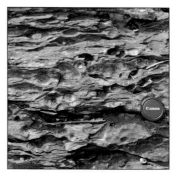

Manganese-rich nodules.

You may find an outcrop where weathering has exposed the dark, uneven top surface of a manganese-rich layer. In these, manganese-rich nodules appear as stained pits and hollows several millimetres wide (see detail).

1000   900   800   700   600   500

Z₁   Z₂   Z₃   €

FYI

- In all, Meguma's turbidites total more than 12 kilometres in exposed thickness. At right, the arrow indicates this site's position in the sequence.

- Depending on circulation patterns, deep water can become starved of oxygen, leading to higher concentrations of certain metals. A well-documented rise in sea level during this part of the Cambrian period likely led to greater water depths over Meguma and to deposition of manganese-rich mud across its sea-floor realm.

- The manganese-rich marker horizon can be recognized even in areas that have been strongly metamorphosed. When heated, the manganese-rich sedimentary rock becomes a rare type of pink, garnet-rich quartzite known as coticule.

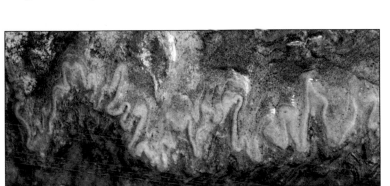

Folded coticule layer north of Sheet Harbour.

Related Outcrops

The manganese-rich marker horizon appears all across the province and signals the beginning of significant changes in the sea covering Meguma. The Cambrian rise in sea level, resulting in deepening water over Meguma, continued to affect sea-water chemistry, as recorded in the next-higher formation (see site 21).

Exploring Further

Maps of magnetic variations in Nova Scotia show the manganese-rich marker as a magnetic anomaly because it also contains significant amounts of iron. These and other anomalies can be used to trace large-scale folds and other regional trends. To view the maps, visit novascotia.ca/natr/meb/geoscience-online/interactive-map-airmag-2DD-map.asp.

500   400   300   200   100   0

€   O   S   D   C   P   Ʀ   J   K   Cz

Alternating layers of black carbon-rich shale and orange, iron-stained siltstone provide a colourful shoreline at The Ovens Nature Park.

New Recipe

Inhospitable Conditions on Meguma's Sea Floor

What has nature cooked up along the shore at The Ovens? The rock layers here look so different from other turbidites that accumulated on the sea floor of ancient Meguma. Without a single shade of grey in sight, the richly coloured layering provides a vibrant background to any shoreline ramble. The explanation for this site's unique palette lies in a key constituent of sea water, oxygen.

As the Cambrian period came to an end, dramatic and sudden change affected the sea covering Meguma. Turbidites continued to accumulate as before, but from end to end of Meguma, conditions on the sea floor became anaerobic—starved of oxygen. In consequence, the sedimentary layers became laden with carbon and sulphur as well as iron, copper, arsenic, and other metals.

The change was probably linked to global changes in sea level, affecting water depth. Eventually oxygen levels rose again, and abundant life forms returned (see site 23). The past was buried as turbidity currents spread the sea floor with fresh sediment from Gondwana's seemingly endless supply. Thanks to later folding, though, nature's kitchen experiment is beautifully presented here for your enjoyment.

Getting There

Driving Directions

Follow Highway 332 between Lunenburg and LaHave to Feltzen South Road and turn (N44.31470 W64.30176) east. Follow Feltzen South Road for about 2.5 kilometres to Ovens Road and turn (N44.32086 W64.27466) east again. Follow Ovens Road to Ovens Natural Park.

Where to Park

Parking Location: N44.32003 W64.25707

Park in the designated area along the clifftop near the park office or as directed by park staff.

Walking Directions

Near the parking area, wooden stairs lead from the clifftop onto the shore. Outcrops begin near the base of the stairs. For additional outcrops, follow a trail south along the shoreline toward the camping area. As conditions allow, cross onto the rock pavement.

Notes

This is a privately run park. For a map of its trail system and for other details, visit www.ovenspark.com.

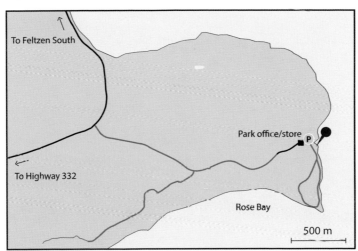

1:50,000 Map

Lunenburg 021A08

Provincial Scenic Route

Lighthouse Route

On the Outcrop

Layers of rust-stained siltstone are interlayered with graphite-rich black slate. The siltstone is stained orange-brown by iron released by the weathering of pyrite and other iron-sulphide minerals.

Outcrop Location: N44.31964 W64.25661

In low water conditions, at the foot of the stairs you can see several low outcrops of characteristic dark rock layers emerging from the beach cobbles. The alternating layers are black and rust-coloured. They have been slightly metamorphosed and folding has tilted them on edge, but the layers originally lay horizontally on the seabed.

The rust-coloured layers are siltstone, each one deposited by a weak turbidity current. When the current passed, the muddy cloud it had carried downslope settled for hours or days afterward. These now take the form of black slate, the dark colour being due to high levels of carbon in the form of graphite.

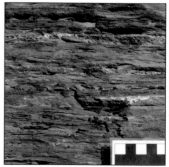

Pitted surface.

Along the shoreline south of the beach, the outcrops have not been worn so smooth. There you can find layers that have become deeply pitted (see detail) as large, cube-shaped crystals of pyrite or arsenopyrite have been weathered, loosened, and washed away.

FYI

- In all, Meguma's turbidites total more than 12 kilometres in exposed thickness. At right, the arrow indicates this site's approximate position in the sequence.

- Physical disturbance of this rock formation (for example, during construction projects) has the potential to create acid rock drainage. As metal-rich, sulphurous minerals are rapidly oxidized by the atmosphere, sulphuric acid is created and the metals are released into the environment.

- To protect the environment and avoid costly damage to infrastructure, since 1995 Nova Scotia has required testing and special precautions for any development affecting sulphur-rich rock types like those in this part of the Halifax Group.

- An underwater state of low oxygen and high sulphur is known as euxinia. The term is derived from the Greek and Latin names for the Black Sea, presently the Earth's largest euxinic sea.

Rust stains on this building at York Redoubt have been caused by weathering of exposed iron-sulphide minerals such as pyrite.

Related Outcrops

You can see examples of the same rock type in a different setting at the York Redoubt National Historical Site of Canada, which overlooks Halifax Harbour from the west. Some buildings there (for example, N44.59698 W63.55310, photo above) were constructed from the iron-rich late Cambrian rock, presumably because it was easily quarried locally. Over time, weathering of abundant iron sulphide minerals has stained the walls yellow, orange, and brown.

By the Way

A white vein of quartz cuts across layers of black slate along the shore at The Ovens.

Meguma's Gold

The beach at The Ovens is a quiet spot these days, but in the summer of 1861 one hardly had room to move. Gold fever had seized Nova Scotia, and The Ovens swarmed with fortune-seekers. By the end of the summer more than 600 hopefuls had arrived to sift through the sand and to break up veins of quartz, looking for coveted nuggets.

In suitable low water conditions you can see signs of what all the fuss was about, in the form of a wide quartz vein in the northwest corner of the low cliff that surrounds the beach (N44.32053 W64.25682). The best-quality gold deposits in Nova Scotia are found in veins similar to this one, formed long after the turbidites of Meguma were deposited.

Given so many years of close scrutiny and prospecting, you can be sure there is very little gold left to be seen at The Ovens now. A glint of yellow may still catch your eye, though. The supply of "fool's gold," or iron sulphide, really is practically unlimited.

FYI

- The beach at The Ovens was originally sandy. The sand was removed in 1861 and 1862 in the process of extracting gold from it.

- Gold-mining districts (shown as yellow dots in the map below) are widely distributed across the Meguma terrane, though they are most numerous north and northeast of Halifax. More than 60 gold-producing districts were active across the Meguma terrane, mostly between 1860 and 1920. During that period the industry produced 1.2 million ounces of gold province-wide.

- The gold occurs mainly in quartz veins deposited from hot fluids that gathered near the top of saddle-like folds. The folding occurred in the late stages of continental collision about 390 million years ago.

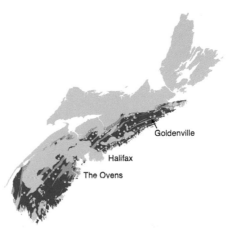

Goldenville

Halifax

The Ovens

Exploring Further

Activities at The Ovens Natural Park include exploration of historic gold mining sites along the Sea Cave Trail and panning for gold. For more information about the park and its programs, visit www.ovenspark.com.

To read "History of The Ovens" (written for the 100th anniversary of the Ovens gold rush), visit the University of Toronto Library archive at archive.org/details/historyofovensst00younuoft.

The Moose River Gold Mines Museum is located on Moose River Road north of Tangier. Moose River Gold Mines Provincial Park (www.novascotiaparks.ca/brochures/Moose_River.pdf) is nearby, on the site of the old mine.

At the Teachers' Centre of the Virtual Museum of Canada, you can view "The History of Gold Mining in Nova Scotia" and "Disaster at Moose River Gold Mine." Visit www.museevirtuel-virtualmuseum.ca/edu/Search.do and enter "Nova Scotia gold" into the search box.

Point Pleasant Park is a popular destination on Halifax Harbour with numerous trails, including Sailors Memorial Way, shown here.

Rocky Ripple
Colourful Turbidites in a Time of Change

The age of the rocks on Point Pleasant places them very near a major division on the geologic time scale, between the Cambrian and Ordovician periods. If Earth history were a novel, this new chapter would be full of surprising developments that change the main characters' lives forever.

All through the Cambrian period, Gondwana and Laurentia moved farther and farther apart as the Iapetus Ocean opened between them. Avalonia, Ganderia, and Meguma all stayed near Gondwana. Early in the Ordovician period, though, the Iapetus Ocean began to close. The new Rheic Ocean opened in parallel to the south, carrying Ganderia and Avalonia away from Gondwana and toward Laurentia (see Geology Basics, The Nova Scotia Story).

For now, Meguma stayed behind on Gondwana's continental margin. With Avalonia drifting away, the sea covering Meguma became part of the rapidly widening Rheic Ocean. Gondwana was now the sole source for Meguma sediment, which long rivers carried farther and farther across the supercontinent's worn surface. Turbidity currents kept tumbling and swirling across the sea floor, but the particles of sediment that reached Meguma became smaller over time.

Getting There

Driving Directions

In Halifax, from the vicinity of St. Mary's University, follow Tower Road south to Point Pleasant Drive. Turn (N44.62615 W63.57445) left (east) and follow Point Pleasant Drive as it curves south and ends in a large parking area by Point Pleasant Park beach.

Where to Park

Parking Location: N44.62520 W63.56380

Park in the parking lot near the beach, on the east side of the park. There is alternative parking near Tower Road on the north side of the park.

Walking Directions

Near the beach, look for signs for Sailors Memorial Way and follow it for about 450 metres to the Canadian Peacetime Sailors Memorial (the Bonaventure Anchor). The outcrop is about 20 metres north of the memorial.

Notes

Dogs are prohibited on Sailors Memorial Way during much of the day. Check park signs or visit www.pointpleasantpark.ca for these and other details.

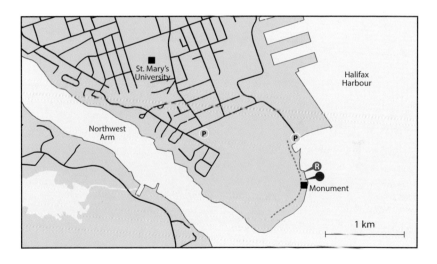

1:50,000 Map

Halifax 011D12

Provincial Scenic Route

Halifax Metro

On the Outcrop

Layers of brown siltstone and black shale alternate to create an eye-catching pattern in this outcrop near the Bonaventure Anchor.

Outcrop Location: N44.62152 W63.56267

In this outcrop, the layers tilt down toward the northwest. On the southeast side (nearest the Bonaventure Anchor), a near-vertical rock face slices down through the layers. There you can see a variety of features in the brown siltstone and black shale.

Each underwater avalanche of watery sediment deposited heavier silt particles first, then mud. If you look closely, within some of the silt layers you can see delicate wisps of a darker colour forming groups of parallel, curved lines. Known as cross-laminations, they formed as the moving current created ripples on the surface of the sediment. (For an example of such ripples, see Related Outcrops.)

High levels of carbon (as graphite) give the shale its black colour, revealing the wavy contacts and variable thicknesses of some layers. Some irregularity could be due to a new turbidity current having scoured channels and hollows in an older mud layer. Also, because mud traps more water than silt does, it later compresses more when the sediment is buried and converted to rock. Some layers of shale were distorted between less yielding siltstone during that process.

FYI

- In all, Meguma's turbidites total more than 12 kilometres in exposed thickness. At right, the arrow indicates this site's approximate position in the sequence.

- Microfossils known as acritarchs in the Halifax Group help to identify the age of the rock layers. They include species typical of Ordovician rocks from cold regions of the southern hemisphere, where Meguma and Gondwana were located at the time.

- Late in the Cambrian period, the sea covering Meguma was starved of oxygen (see site 21). By the time the rocks here formed, oxygen levels had increased, perhaps because formation of the Rheic Ocean opened the sea to new, oxygen-rich currents.

km

12

10

8

6

4

2

0

Halifax

Goldenville

Related Outcrops

From the main outcrop, walk about 100 metres along Sailors Memorial Way toward the beach and parking lot to find a second outcrop (N44.62233 W63.56315). Look for the northwest-slanting top surface of the rock layers. There, you can see ripples in siltstone, created by a turbidity current and preserved under a subsequent layer of mud.

Ripple marks on a tilted siltstone layer.

The layers at Point Pleasant are part of the Halifax Group (shaded light blue in the map above), a 4-kilometre-thick succession of fine-grained turbidites (also see sites 21, 23, and 24).

Sedimentary layers are gently arched and tilted along the shore at Blue Rocks, near Lunenburg.

Portrait of a Fold

Sedimentation and Tectonics Sculpt an Outcrop

A cultural transformation took hold in the quiet fishing village of Blue Rocks in the 1930s. It was discovered by artists "from away." Britain's Stanley Royle and New England's Marsden Hartley were among the painters who found their way here in the early years. The community's artistic lineage continues today with numerous artists in residence.

And no wonder: From the scenic shore with its fishing sheds and craft, to the historic St. Barnabas church, Blue Rocks offers a visual feast. Even its rocky foundation fascinates the eye. With a subtle palette of greys, browns, and blues, nature has rendered a complex portrait of Meguma in the uniquely beautiful outcrops along the shore.

The layers of silt and mud preserved at this site were deposited on the edge of the young Rheic Ocean. Sediment reaching Meguma from Gondwana had become scanty. Each turbidity current delivered just a few thin layers of silt followed by thick mud. Millions of years later, in the final stages of the ocean's closing, the layers were folded. The rock's arresting texture resulted from the interplay of large- and small-scale deformation.

Getting There

Driving Directions

From Highway 332 on the east side of Lunenburg, watch for the intersection with Blue Rocks Road. Turn (N44.36975 W64.29047) east and follow Blue Rocks Road for about 3.8 kilometres. After passing Blue Rocks Pond on the left, watch for a pull-off on the right, along the shore.

Where to Park

Parking Location: N44.35622 W64.24880

Park in the gravel pull-off next to the boulder barrier.

Walking Directions

From the pull-off, make your way across the boulder barrier and down onto the outcrop. In low tide conditions, you can explore more extensively along the rocky shoreline.

Notes

This site is in a residential area and the road is not wide. Please be considerate of traffic and of residents' privacy and property. Park only in the pull-off.

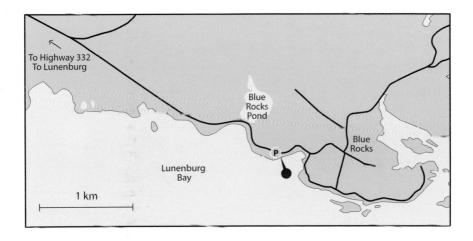

1:50,000 Map

Lunenburg 021A08

Provincial Scenic Route

Lighthouse Route

On the Outcrop

Nearly horizontal sedimentary layers are arched and crumpled into folds of various sizes. A strong pattern of vertical cleavage intersects the layers.

Outcrop Location: N44.35608 W64.24851

The sedimentary layering here is approximately horizontal because the site is located in the crest of an anticline a few kilometres wide. Folds of that size can be seen only on a geologic map or aerial photograph. However, smaller-scale features related to the anticline are visible here.

Look for folds on a variety of scales, using colour contrasts in the sedimentary layering as your guide. The thinner layers striped with shades of light grey were deposited as silt; the thick, dark blue-grey layers were originally mud. You may find some broad arches several metres across and others measuring tens of centimetres from crest to crest.

Looking closer, you may notice that colour banding in some individual siltstone layers traces folds just a few centimetres wide. During deformation the adjacent shale was ductile, behaving like paste. When the siltstone layers crumpled, the shale squeezed into the narrow v-shaped spaces between the folds. Superimposed on all the folds is a near-vertical cleavage. Combined with weathering, these structures account for the beautiful flame-stitch pattern in the rocks.

FYI

- In all, Meguma's turbidites total more than 12 kilometres in exposed thickness. At right, the arrow indicates this site's approximate position in the sequence.

- At this site, you may also find thicker siltstone layers with conspicuous cross-laminations as well as large, cube-shaped crystals of rusty-looking pyrite.

- Folding can tilt and buckle sedimentary layering into any orientation, but cleavage always forms perpendicular to the direction in which the rock has been squeezed (see diagram).

Pyrite in siltstone.

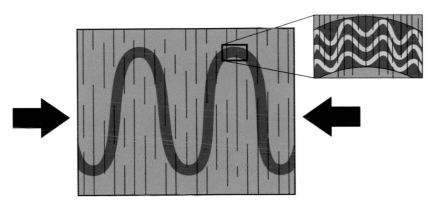

Compression in the crest of a fold can cause vertical cleavage and small-scale crumpling of layers.

Related Outcrops

Because the Goldenville and Halifax groups were folded together into complex shapes during the Devonian period, most outcrops of Meguma's turbidites now have cleavage. At sites 19 and 21, located in the sides, or limbs, of a fold, the cleavage is almost parallel to the sedimentary layers. At site 17, as here, located in the nose, or hinge, of a fold, it is nearly perpendicular.

500	400	300	200	100	0
Є O S	D	C	P	Ŧ J	K Cz

Sun burns off the morning fog at Cape St. Marys lighthouse—Nova Scotia's westernmost mainland beacon—near the site of a significant unconformity.

Old and New
An Eventful Gap in Meguma's Rock Record

Silt—check, mud—check ... silt—check, mud—check. As seen at sites 22 and 23, in some ways the rock layers of early Ordovician Meguma resemble the inventory of a vast sediment warehouse. As long as Meguma remained attached to Gondwana, the results were quite predictable. At Cape St. Marys, though, you'll find something new in stock.

Throughout the rest of the Ordovician period, the arrangement of Earth's tectonic plates changed rapidly and dramatically. The Rheic Ocean grew thousands of miles wide, creating a complex, ever-changing collage of ocean tracts and microcontinents. By the time Ganderia collided with Laurentia early in the Silurian period (see sites 13 and 14), Meguma had begun to rift away from Gondwana at last.

At Cape St. Marys you can see both the old and the new environment of Meguma side by side. Between them is a fine line, an unconformity, representing a 30-million-year gap in Earth's record-keeping. On one side, the slate warehouse; on the other, Meguma's innovation: layers of much younger volcanic ash. Explosive eruptions marked the onset of new conditions.

Getting There

Driving Directions

Along Highway 1 in Mavillette watch for signs for Cape St. Marys. Turn (N44.10087 W66.18518) southwest onto Cape St. Marys Road and follow it for 2.8 kilometres past Mavillette Beach and through the community of Cape St. Marys. Where the paved road ends, fork left to the public wharf.

Where to Park

Parking Location: N44.08447 W66.20750

Park in the gravel parking area at the public wharf.

Walking Directions

From the parking area, walk back up to the main road and turn left. Follow the gravel road to the shore, then in low tide conditions make your way down the boulder barrier to a small gravel beach. Cross the beach and follow one of the footpaths up onto a grassy, rocky knoll. From the knoll you can view and explore two rock types and the contact between them.

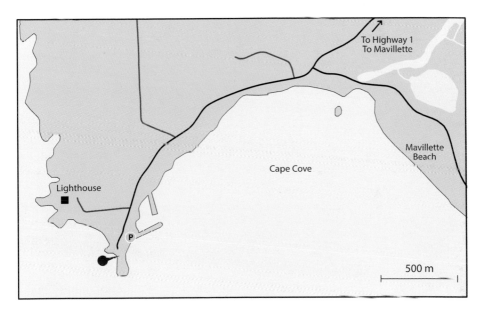

1:50,000 Map

Meteghan 021B01

Provincial Scenic Route

Evangeline Trail

On the Outcrop

(a) Dark grey slate on the right contrasts with pinkish- and greenish-grey volcanic rocks on the left along the unconformity: (b) tuff with visible grains of quartz and feldspar; (c) slate with vertical cleavage and slightly tilted sedimentary layering.

Outcrop Location: N44.08334 W66.20849

From the grassy knoll or ridge on the beach, face seaward. In low water conditions, look for dark grey slate on the right. Once you see an example of slate, look at the next rock to the left of it. Still slate? Look farther to the left.

You'll see that one part of the shore in front of you looks slightly hollowed out and perhaps sandier. There, the rock is golden brown and broken into irregular vertical tablets or slabs. Farther left still, the colour changes to pinkish grey, then eventually greenish gray. These rocks are all known as tuff, formed when ash, crystals, and rock fragments were erupted from a volcano.

The contact between the slate and the tuff is the unconformity that makes this site so significant. The rocks right beside it are broken and weathered because later on, a shear zone formed along the unconformity. All the rocks here—both slate and tuff—were metamorphosed and deformed during the Devonian period. They all have well-defined, near-vertical cleavage as a result.

1000   900   800   700   600   500

Z₁   Z₂   Z₃   €

FYI

- Chemical elements we rarely think about help piece together the geologic past. Measuring the amount of zirconium, niobium, vanadium, and yttrium in a volcanic rock can help pinpoint whether the volcano was part of an oceanic ridge or island, a mountain range, or a continental rift valley. Based on their chemical characteristics, the volcanic rocks here formed during continental rifting.

- In contrast to the deepwater environment of Meguma's turbidites (see sites 17–23), the volcanic ash here seems to have fallen sometimes into shallow water and sometimes onto dry land.

- Despite later metamorphism and deformation, two kinds of tuff can be easily distinguished here (photo below). In terms of minerals and rock chemistry, the pale grey tuff is similar to rhyolite, and the greenish-grey tuff is similar to basalt.

Tuff with characteristics similar to rhyolite (left) and basalt (right) at Cape St. Marys.

Related Outcrops

The volcanic rocks here are among the only rocks of Silurian age in Meguma. They are part of the Rockville Notch Group (shaded blue in the map at right).

You can explore a much thicker, well-preserved sequence of related volcanic layers at Cape Forchu (site 25).

The slate here at Cape St. Marys is equivalent in age to the slate at sites 22 and 23.

500   400   300   200   100   0

Є O S D C P Ŧ J K Cz

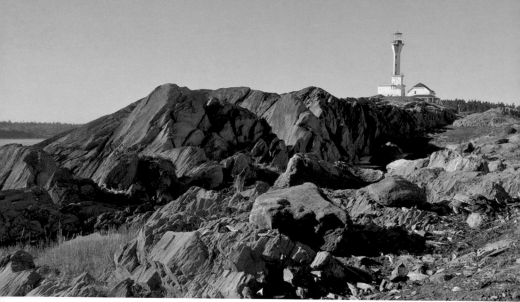

Steeply tilted layers of grey tuff dominate the headland at Cape Forchu lighthouse on Yarmouth Sound.

Fiery Fragments
Thick Volcanic Ash Deposits of Meguma

The Cape Forchu lighthouse has been witness to more than 150 years of maritime history as ships under sail and steam have passed by on Yarmouth Sound. Standing near the lighthouse today, it's easy to imagine the heyday of ocean travel, when colourful flags and funnel markings announced the origins of seafaring traffic from near and far.

The rocks that are such a prominent feature of the headland signal their origins, too, and the news is remarkable. During the Silurian period, after 100 million years of relative quiet, Meguma became the site of explosive volcanic activity known as pyroclastic eruptions. This term comes from two Greek words meaning "fiery fragment": Time and time again, volcanic ash and other hot debris were ejected into the air. They fell in layer after layer to form a rock type known as tuff.

At this site, the ash and other volcanic debris fell into shallow water. Although the rock layers were metamorphosed and tilted long afterward, many of their original features are well-preserved. They include volcanic "bombs," which were erupted as blobs of molten rock that solidified before reaching the ground.

Getting There

Driving Directions

Highway 304 (Vancouver Street) on the west side of Yarmouth passes the Yarmouth Regional Hospital, then follows Grove Road and turns sharply left at Yarmouth Bar Road. From that point, follow Highway 304 south about 7.2 kilometres farther—it crosses two causeways before terminating at the Cape Forchu lightstation.

Where to Park

Parking Location: N43.79502 W66.15424

This is one of the designated parking areas at the site.

Walking Directions

From the parking area follow the trail system south along the low ground beside the lighthouse, choosing the path that most closely follows the long rocky ridge toward the point. There are outcrops along the path. For additional features of interest, at a convenient spot (for example, N43.79223 W66.15522) cross onto the rock surface to continue south along the ridge.

Notes

This site is owned and maintained by the Municipality of the District of Yarmouth, with the participation and support of the Friends of the Yarmouth Light Society. For more information visit www.capeforchulight.com.

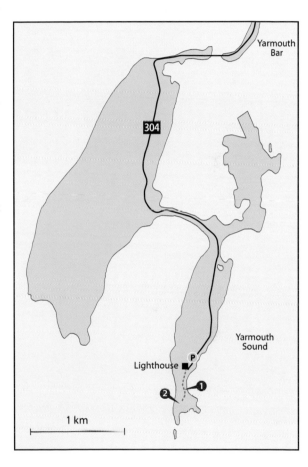

1:50,000 Map

Yarmouth 020O16

Provincial Scenic Route

Evangeline Trail

On the Outcrop

(**a**) At the main outcrop location, steeply tilted layers of volcanic ash preserve a record of numerous eruptions. (**b**) In a rock face farther south (see text), nearly vertical layers contain numerous volcanic bombs. Note that "up" is to the left.

Outcrop Location: N43.79289 W66.15505

At this site along the trail, the steeply tilted layers, or beds, of ash are emphasized by weathering of the outcrop. It's easy to see the subtle variations in texture and colour that mark different pulses of ash (photo **a**). Everywhere on the headland, as you face south, older layers are on the right and younger layers on the left.

The surface of many layers is lumpy, due to the irregular nature of the pyroclastic debris. In some thicker layers, you may see (and sometimes feel) what is known as graded bedding—with a coarse texture at the bottom, gradually changing to fine at the top. It is characteristic of ash that settled in water.

In some locations on the headland (for example, along the rocky ridge about 250 metres south of the lighthouse; N43.79187 W66.15558) are volcanic bombs. The site is well-known among geologists for a feature known as "bomb sag" (photo **b**). It was created when a large fragment fell into still-soft ash, pressing the layers down when it landed. This kind of evidence helps geologists recognize the layers' original orientation.

FYI

- The rocks here are part of a pile of volcanic material as much as 10 kilometres thick. Analyses reveal little variation in chemical characteristics from bottom to top, which suggests the entire volume was erupted in a relatively short period of time, perhaps just one or two million years.

- Based on chemical characteristics, geologists think these volcanic rocks formed in a continental rift, somewhat like the East African Rift. However, the rift in Meguma was mainly under water, as shown by the well-developed layering characteristic of water-lain volcanic ash.

- During the Silurian period, while Meguma was just beginning to separate from Gondwana, Avalonia was drifting toward Laurentia (see site 5). Ganderia had already reached Laurentia, and a collision was under way (see sites 13 and 14).

Volcanic rocks form the backdrop to inviting amenities at Cape Forchu Lighthouse Park.

Related Outcrops

The rocks at this site are equivalent to the darker, greenish-grey volcanic layers at Cape St. Marys (site 24).

On Cape Breton Island you can see tuff containing volcanic bombs and other pyroclastic debris at Louisbourg lighthouse (site 9). It formed much longer ago as part of the microcontinent Avalonia and is not related to the volcanic activity preserved at Cape Forchu.

A grass-covered gravel track follows the shoreline toward outcrops near the wind farm on Pubnico Point.

Heating Up
Unusual Minerals Formed during Metamorphism

Mud has such humble beginnings. Nothing could be more mundane or commonplace. Meguma's huge thickness of turbidites certainly had plenty of it (see sites 20–23). Most layers of mud deposited on Meguma's sea floor now take the form of shale or slate, but at this site it has been thoroughly transformed into a more flamboyant rock.

Mud is geologically interesting because it has distinctive chemical characteristics—for example, generous amounts of aluminum—as compared to silt or sand. When the heat and pressure of metamorphism affect sedimentary rock, a wider range of minerals can form in mud-derived rock than in the other common types. At this location, metamorphism of mudstone caused the growth of an attractive mineral, andalusite.

The andalusite crystals here are so uniform in size and shape they almost seem manufactured. Some of the outcrops look like a truckload of the crystals has spilled on the shore. Their slightly pink hue and speckled texture give the site a festive air. It's rare to see such a high concentration of andalusite, giving mineral enthusiasts who visit the site a cause for celebration.

Getting There

Driving Directions

From Highway 103 or Highway 3 between Yarmouth and Shelburne, follow Highway 335 south for about 13.5 kilometres through West Pubnico and Middle then Lower West Pubnico, all the way to the wind farm on Pubnico Point. Continue past rows of windmills to the parking area.

Where to Park

Parking Location: N43.59660 W65.79800

This is an open gravel area by the gate for one of the wind-farm access roads. Please do not block the gate.

Walking Directions

From the parking location, continue southwest on the extension of Route 335. First as a gravel road and then as a grassy, cobbled, or earthen track it follows the shoreline out to the point, providing many access points to rock outcrops on the shore.

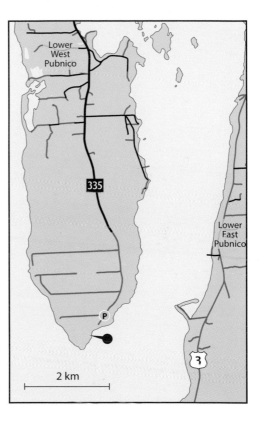

1:50,000 Map

Pubnico 020P12

Provincial Scenic Route

Lighthouse Route

On the Outcrop

Outcrops along the shore are crowded with long, chunky crystals of andalusite. Other mineral grains inside the andalusite give the large crystals a speckled appearance.

Outcrop Location: N43.59349 W65.80087

This and nearby outcrops contain abundant crystals of andalusite. Typically 3 to 5 centimetres long, these rod-shaped crystals resist weathering, giving the outcrop a bumpy appearance. The crystals are not aligned, instead looking as though they had been scattered randomly in the rock. Some areas are nearly 50 per cent andalusite.

In wave-washed outcrops or boulders you can see other minerals in the rock more clearly. Along with andalusite, white quartz, feldspar, and muscovite give the rock its light colour. You may find examples in which the pale pink colour of the andalusite stands out from the white background of the other minerals. In this pale matrix are abundant flecks and chunks of biotite and, less commonly, brown garnet and staurolite.

As andalusite crystals grew during metamorphism, they engulfed existing smaller crystals of quartz, biotite, garnet, and other minerals. Large crystals with this type of texture are called poikiloblasts.

1000   900   800   700   600   500

Z₁   Z₂   Z₃   €

FYI

- Andalusite is one of a group of minerals made of aluminum and silica (Al_2SiO_5). In some places (but not in Nova Scotia) it is quarried to make ceramic brick for use in high-temperature processes such as steel manufacturing.

- The entire Meguma terrane was metamorphosed during the Neoacadian orogeny (see site 27), but metamorphic grade varies across the province. In the map at right, the colour scale yellow-orange-red-violet shows increasing metamorphic conditions, ranging from low grade in central parts of Nova Scotia to high grade in the east and especially in the southwest.

Related Outcrops

In this and other outcrops along the shore at Pubnico Point, narrow veins of andalusite can be seen cutting across the rock pavement. According to some geologists, this is impossible because aluminum (necessary for formation of andalusite) is not readily carried in fluids. Yet here they are, their origin as yet unexplained.

The rock at this site is part of the same manganese-rich rock formation seen at Rainbow Haven (site 20), where it is much less metamorphosed. Its mud-rich layers were transformed into the andalusite-rich zones seen here.

A narrow vein of andalusite (see arrow) cuts from lower left to middle right in this photo from Pubnico Point.

Shoreline outcrops near the picnic facilities and grounds of the Sandy Point Lighthouse Community Centre are easily accessible.

Telltale Mineral

Staurolite Schist Recording High Temperatures

About 390 million years ago, a major mountain-building event, the Neoacadian orogeny, affected the Meguma terrane. It folded and metamorphosed all of the microcontinent's sedimentary and volcanic rocks, leaving clear indications of a significant plate collision. However, nothing similar happened in Avalonia, Meguma's present-day neighbour. Details of the collision remain a mystery.

Another enigma is the pattern of metamorphism (see site 26, FYI). In southwestern Nova Scotia, temperatures appear to have been quite high, but pressures were moderate. In normal continental conditions, temperature and pressure increase together at deeper and deeper levels of the crust. In this part of Meguma, an unusual amount of heat reached shallow levels of the crust.

Of course, we can't travel deep into the Earth's crust, much less back in time. Geologists rely on metamorphic minerals to capture a record of past conditions. The schist at this site contains typical minerals such as quartz, feldspar, biotite, and garnet, but also the attractive and informative mineral, staurolite. Its presence indicates that temperatures in this area reached 500°C at a depth of less than 10 kilometres.

Getting There

Driving Directions

From Highway 3 (King Street) in Shelburne, turn (N43.76392 W65.32138) south onto Hammond Street. South of Shelburne near the Roseway Hospital, Hammond Street becomes Sandy Point Road. From there, continue for about 6.8 kilometres. Watch for a large sign at the Sandy Point Lighthouse Community Centre and turn right into the park entrance.

Where to Park

Parking Location: N43.69169 W65.32330

Park in the gravel area by the community centre.

Walking Directions

From the parking area, walk west toward the shore. Beside a wooden viewing platform use the ramp to access the beach. In low tide conditions walk right (north) along the shore to outcrops near the north side of the community centre building, that is, on the shore in front of the boardwalk.

Notes

The Sandy Point lighthouse is designated a Municipal Heritage Property.

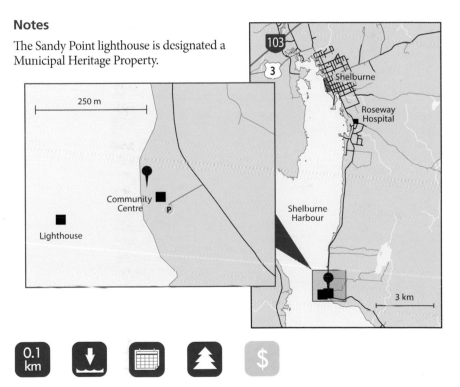

| 0.1 km | | | | $ |

1:50,000 Map

Lockeport 020P11

Provincial Scenic Route

Lighthouse Route

On the Outcrop

As the tide falls, you can find low outcrops of schist among sand and boulders along the shore in front of the community centre.

Outcrop Location: N43.69200 W65.32372

The low outcrops are grey in overall colour. Some surfaces are stained brown due to the iron oxide content of the rock. Variation in the amounts of light and dark minerals gives the rock a banded appearance as well as a coarse cleavage along which the rock tends to break. Both these features are generally parallel to the shoreline.

Looking closely you can see that the white background of quartz and feldspar is finely speckled with biotite, a dark mica. The rock also contains small, round, red garnet crystals. In many outcrops the largest crystals are staurolite (see detail). These reddish-brown crystals are angular, with sharply defined edges. Some look like the diamond from a deck of cards.

Staurolite crystals in schist.

Because of the mineral's shape, hardness, and resistance to weathering, outcrops containing staurolite often feel very rough. If you're not sure whether the reddish mineral you see is garnet or staurolite, your sense of touch may provide a clue.

1000		900		800		700		600		500
	Z_1			Z_2				Z_3		€

FYI

- Staurolite sometimes forms interlocking crystals in the shape of a cross, an "X", or similar arrangements. This happens due to a process known as twinning. The name *staurolite* refers to this phenomenon, being derived from two Greek words meaning "star rock."

- In appropriate low tide conditions, it is possible to walk out to the lighthouse. Near it, you may find additional outcrops of staurolite schist, some cut by veins also containing staurolite, biotite, garnet, and quartz.

- In some outcrops and boulders along the shore you may see dark veins cutting the schist. Some are no wider than a pencil stroke, but a few are several centimetres wide. The veins look like they are filled with black resin, but they are in fact a rare rock type—pseudotachylite. It forms by instantaneous melting when a rock is shocked by a meteorite impact or earthquake. Pseudotachylite has the same composition as the surrounding rock—its dark colour is due to a glassy texture.

Pseudotachylite vein in schist.

Sandy Point Community Centre.

Related Outcrops

Prior to metamorphism, the rocks at this site would have resembled those at Rainbow Haven (site 20).

Related features of interest in southwestern Nova Scotia include a rare and attractive andalusite schist (see site 26) and migmatite, a rock that became so hot it began to melt (see site 28).

00   400   300   200   100   0

€   O   S   D   C   P   Ṛ   J   K   Cz

Accessible via a sandbar at low tide, this outcrop of migmatite provides a conspicuous landmark on the long beach at The Hawk, Cape Sable Island.

Tectonic Sauna
Partial Melting as Hot Rock Gets Hotter

The beach is spectacular and seemingly endless, home to dozens of seabird species. Nearby Cape Sable lighthouse is Nova Scotia's tallest, more than 30 metres high. And the outcrop ... well, its appearance may be modest in comparison to its surroundings, but the rocks themselves are beautiful. They provide a glimpse of metamorphic conditions seen at just a few locations in the province, conditions so extreme they caused parts of the rock to melt.

Rock that was once partly solid and partly molten is known as migmatite, a name derived from a Greek word meaning "mixture." Sometimes migmatite forms across an entire region, for example, if plate collision pushes continental crust to great depths. In this case, though, the effect was localized. Meguma's sedimentary rocks were metamorphosed but still solid until a nearby igneous intrusion heated them further.

Each mineral has a specific temperature at which it begins to melt. For quartz and feldspar, the threshold is relatively low, about 600° to 750°C. Because they are made of quartz and feldspar, the once-molten portions of the rock are light in colour, making it easy to see what happened.

Getting There

Driving Directions

From Highway 3 southwest of Barrington Passage, turn (N43.52138 W65.61610) south onto Highway 330 and cross onto Cape Sable Island. Follow Highway 330 through Clarks Harbour. Watch for Hawk Point Road and turn (N43.43178 W65.61618) right (south), following it for about 1.7 kilometres to the intersection with New Road. Turn (N43.41780 W65.62097) left (east) onto New Road and follow it to the shore.

Where to Park

Parking Location: N43.41618 W65.61431

Park in the open area where New Road ends at the shore.

Walking Directions

From the parking area, walk east onto the beach. Turn left (north) and walk about 100 metres along the shore. There, in low tide conditions, a sandbar connects the main beach with a rocky prominence about 50 metres to the east. Cross onto the rocky area to view the outcrop.

Notes

The beach at this site is designated a Protected Beach under provincial law.

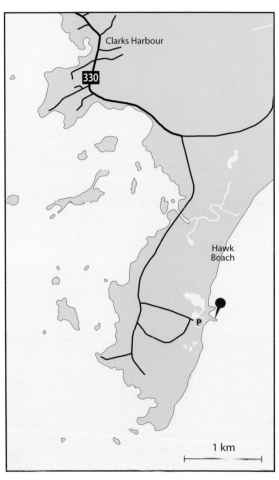

1:50,000 Map

Cape Sable Island 020P05

Provincial Scenic Route

Lighthouse Route

On the Outcrop

Broken boulders on the outcrop provide a detailed look at the effects of partial melting. Here, fine, millimetre-scale layering is cut by larger, irregular veins of quartz and feldspar.

Outcrop Location: N43.41658 W65.61241

The best way to see what happened to this rock is to look for fractured boulders or other broken, fresh surfaces. Because light-coloured minerals melt most readily, and because the melted material tends to gather together, migmatite consists of distinct light and dark areas. Here the process took place on several different scales. Thin layers of melt, just millimetres wide (see detail), sweated out of the rock as the temperature rose. In addition, larger amounts of melt were squeezed into zones of weakness, forming irregular, cross-cutting veins.

Layering in migmatite.

All the while as melting took place, the rock was being deformed by tectonic forces acting on the region. High temperatures and partial melting both weakened the rock. For that reason, you'll see many folds, kinks, and swirls in the layering.

FYI

- Partial melting at The Hawk was triggered by the Barrington Passage tonalite, which underlies the wide peninsula between Upper Woods Harbour and Barrington Head on the mainland. The tonalite is about the same age as the granite at Black Duck Cove (site 30).

- Many Devonian intrusions in Meguma did not cause melting in surrounding rocks. Most were granite magma (which is less hot than tonalite magma) and intruded rocks at a lower metamorphic grade, so the required temperatures were not reached.

- Unlike the light minerals that melted here, dark minerals such as biotite and amphibole resist melting unless the temperature gets much higher, about 1000° to 1200°C. The dark portions of the migmatite remained solid.

Related Outcrops

Before they were metamorphosed, the rocks at The Hawk were turbidites similar to those farther east, for example, at Green Bay (site 19). See sites 26 and 27 for more information about Devonian metamorphism in the region.

Outcrops and boulders line the scenic shore near Summerville.

Features at the site include migmatite and pegmatite.

To see another migmatite formed near an igneous intrusion, follow Highway 3 about 1.4 kilometres east from the entrance to Summerville Beach Provincial Park. From a pull-off (N43.94978 W64.79902) on the south side of the road, follow a path up over a wooded knoll. As conditions allow, climb onto the outcrop to view the rock pavement or view it from the knoll.

500   400   300   200   100   0

Є   O   S   D   C   P   Ṟ   J   K   Cz

The iconic lighthouse at Peggys Cove sits on a glacially sculpted outcrop of Devonian granite.

Sea of Granite
Crustal Melting beneath Meguma

Peggys Cove is a popular destination. The lighthouse and rocky headland have graced the pages of innumerable books, pamphlets, posters, and paintings, inspiring and welcoming visitors to the province. To the list of its attractions, you can add the rock on which you stand to admire the famous view. Its history is awe-inspiring in its own way.

The outcrops here are all granite—part of a huge mass of intrusions known as the South Mountain Batholith. In fact, as you walk the rounded surfaces at Peggys Cove, you are in contact with the largest body of granite in the Appalachians. It extends in a broad arc northwest from Halifax to the Annapolis Valley then southwest toward Yarmouth (see Related Outcrops).

When the granite formed, Meguma was sliding against Avalonia to the north; Gondwana was approaching from the south as the Rheic Ocean closed. The granite shows no signs of collision, no mineral alignments suggesting deformation. On the outcrop, pause to consider the batholith, larger than Prince Edward Island, accumulating in the crust near the end of Meguma's journey.

Getting There

Driving Directions

From Highway 3 in Upper Tantallon, follow Highway 333 south. About 4.2 kilometres southeast of Indian Harbour, watch for Peggys Point Road on the right. Fork right (N44.49772 W63.91528) and follow the road through the community of Peggys Cove. The road ends at the lighthouse.

Where to Park

Parking Location: N44.49196 W63.91698

This is the parking lot for a large restaurant and gift shop.

Walking Directions

From the parking area, you can follow any of several footpaths and sidewalks onto the granite outcrops around the lighthouse.

Notes

Although outcrops are accessible in normal tide conditions, high surf can create dangerous conditions along the shore.

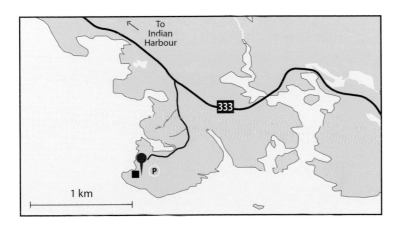

1:50,000 Map

Sambro 011D05

Provincial Scenic Route

Lighthouse Route

On the Outcrop

(**a**) Extensive outcrops near the lighthouse hint at the huge volume of granite that intruded Meguma. (**b**) Large, rectangular crystals of potassium feldspar make up about one-third of the rock.

Outcrop Location: N44.49177 W63.91833

Note that the waypoint provided is near the lighthouse, but the granite can be examined from many possible vantage points on the rocky headland (photo **a**).

Three minerals are found in roughly equal proportions in the rock. Potassium feldspar is most conspicuous, typically appearing as large, rectangular crystals in a finer-grained matrix (photo **b**). The other primary minerals are quartz (grey) and a second feldspar, plagioclase (light grey to white). Their shape is more irregular. Giving the rock a speckled appearance is the dark mica, biotite. Muscovite, a white mica, is also present but less noticeable.

Cutting across some outcrops are narrow dykes of aplite and pegmatite. They contain minerals similar to those in the granite. Aplite has a fine-grained, sugary texture and lacks biotite, while in pegmatite the crystals are unusually large.

In some locations you may also find fragments of dark grey rock surrounded by granite. These are bits of the rock into which the granite intruded—Meguma's metamorphosed sedimentary rocks.

1000   900   800   700   600   5(

Z_1 Z_2 Z_3 €

FYI

- The granite at Peggys Cove is part of the Halifax pluton, which underlies nearly the whole peninsula between Halifax Harbour and St. Margarets Bay. About a dozen plutons make up the South Mountain Batholith, which has a total exposed area of 7,200 square kilometres and an estimated total volume of 100,000 cubic kilometres.

- The granite of South Mountain Batholith is an important source of aggregate for highway construction and is also quarried for ornamental use. Beautiful examples of the granite used as monument stone can be seen at the Swissair Memorial Site about 1 kilometre northwest of Peggys Cove.

- Granite magma can form by melting of continental crust dragged deep into the Earth by plate tectonic processes. Yet the batholith's chemical characteristics show that its plutons did not form by melting of Meguma's thick turbidite sequence. Their deep crustal source remains uncertain.

- During intrusion, a lot of metamorphosed sedimentary rock was caught up in the granite magma, contaminating it with unusual amounts of aluminum. For that reason in some locations the batholith contains minerals not typically found in granite, including garnet, andalusite, and cordierite.

Related Outcrops

Intrusions of Devonian granite (shaded blue in the map at right) are widespread in the Meguma terrane of Nova Scotia. They include the large South Mountain Batholith (SMB) as well as smaller bodies of granite seen, for example, near Shelburne (S), Port Mouton and Kejimkujik National Park—Seaside (PM), Musquodoboit (M), and Canso (C).

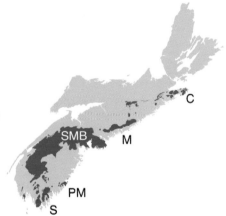

For an example of one of Meguma's smaller Devonian granites, see site 30, Black Duck Cove.

For an unrelated Devonian granite of Ganderia containing abundant examples of aplite and pegmatite, see site 15, Green Cove.

500   400   300   200   100   0

€   O   S   D   C   P   Ṫ   J   K   Cz

No matter the weather, the trails at Black Duck Cove Provincial Park near Dover offer fine scenery, amenities, and access to granite outcrops.

Nature's Quarry
One of Meguma's Smaller Devonian Granites

In Black Duck Cove Provincial Park, as you walk along the trail in the popular area known as Flat Rocks, you may see large blocks of granite in the grass and brush on the landward side of the trail. They were tossed there, as local residents can tell you, by Hurricane Juan in 2003. And there they sit to this day, providing silent commentary on the titanic forces of nature.

The granite outcrops on the other side of the trail have a similar message of nature's immensity, though longer ago and deep within the Earth. As the Rheic Ocean closed late in the Devonian period, a subduction zone beneath Meguma became a granite-generating behemoth from which magma rose during a period of 15 or perhaps even 20 million years.

In addition to the huge bulk of the South Mountain Batholith (see site 29), several smaller bodies of Devonian granite are exposed along Nova Scotia's Atlantic shore from Barrington Passage to Dover Bay. The handsome white granite at Flat Rocks was part of that long-lived and widespread event.

Getting There

Driving Directions

Along Highway 16 about 3.5 kilometres west of Canso, watch for Dover Road and turn (N45.32497 W61.03429) south. Follow Dover Road almost 7 kilometres, continuing through the community of Dover to a recreation area with a baseball diamond and other amenities. Turn (N45.27672 W61.03266) left (east) at the entrance to Black Duck Cove Provincial Park and continue into the parking area.

Where to Park

Parking Location: N45.27654 W61.03143

Park in the gravel lot just beyond the park entrance.

Walking Directions

From the parking area, walk around the left (east) side of the park building and follow the sequence of trails and boardwalks past the beach and around the point to the area called Flat Rocks.

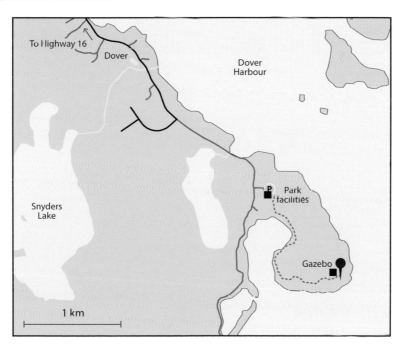

1:50,000 Map

Chedabucto Bay 011F06

Provincial Scenic Route

Marine Drive

On the Outcrop

Blocks and slabs of granite along the shore at Flat Rocks are not quarried, but formed by the natural process of exfoliation.

Outcrop Location: N45.27083 W61.02508

The area along the trail known as Flat Rocks is aptly named. It looks like the remains of a quarry, with large, flat slabs and blocks of granite along the shore. Not man-made, the fragmentation is caused by a natural process known as exfoliation. This complex process causes cracks to form parallel to the surface of very homogeneous rocks, including granite.

The broken surfaces provide plenty of opportunity to see the even texture of the rock. The minerals are those typical of granite: patches of grey quartz, potassium feldspar with a slightly pinkish cast, white plagioclase feldspar, and black biotite.

(a) Pegmatite in granite; (b) pegmatite detail, showing tourmaline.

Less common features here are veins and irregular pods of pegmatite (photo **a**) containing the mineral tourmaline. A semi-precious mineral, it is most easily recognized when it appears as black, rod-like prisms among lighter minerals (photo **b**).

1000	900	800	700	600	50(
Z₁		Z₂		Z₃	€

FYI

- The end of the Devonian period was very eventful for the Meguma terrane. The granite plutons were intruded between 380 and 360 million years ago at a depth of about 10–12 kilometres in the crust, but afterward the terrane was uplifted very rapidly. By 360 million years ago, some of the plutons were already being eroded at the Earth's surface, as shown by the presence of granite fragments in late Devonian sedimentary rock.

Related Outcrops

See site 29, FYI, for a map showing the distribution of Meguma's Devonian granites.

By the Way

About 10 to 12 kilometres southeast of Guysborough, Highway 16 follows the unusually straight shore of Chedabucto Bay (see map). A steep ridge rises sharply along the south side of the highway. To the north, the land drops steeply to the waters of the bay.

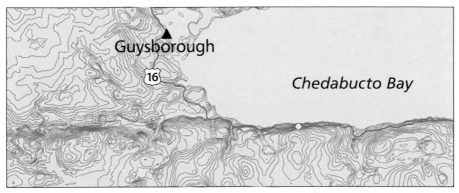

Highway 16 and the Chedabucto fault trace near Guysborough (yellow dot marks pull-off).

You can stop at a pull-off along the way (N45.34856 W61.41356) to examine this break in topography. Appearing in the map above as an east–west line of closely spaced topographic contours, it is the trace of the Cobequid-Chedabucto fault. This fault and its many branches bisect Nova Scotia, marking the approximate northern boundary of the Meguma terrane against Avalonia and the site of their oblique collision. For more information about the fault and its appearance along the Bay of Fundy, see site 31.

An outcrop by a pull-off provides easy access to the Cobequid fault scarp along Highway 209 west of Parrsboro.

Strike-Slip
Signs of Movement on the Cobequid Fault

Astronauts can see the whole thing through the window of the International Space Station … and you can see part of it through the window of your car along Highway 209 west of Parrsboro. Why not get out for a closer encounter with the Cobequid-Chedabucto fault system? It's a major break in the Earth's crust, separating the Meguma terrane from Avalonia.

When Meguma approached Avalonia, it wasn't a head-on collision. Instead, Meguma sideswiped its new neighbour, sliding from what is now the east, dragging and scraping against Avalonia for several million years. This type of interaction is known as strike-slip, or transcurrent, motion. The San Andreas fault system slicing through southern California provides a well-known modern example.

Once plate collision created a zone of weakness in the crust, later stresses caused new displacements—throughout the Carboniferous as Gondwana collided with Meguma and perhaps even later as Pangaea broke apart. For that reason, the age of the rock at this site and of all the movements affecting it are uncertain. What is certain is the cause of the weakness: Meguma's arrival.

Getting There

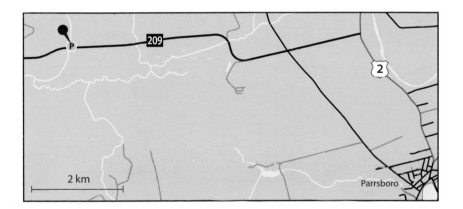

Driving Directions

From Highway 2 on the north side of Parrsboro, turn (N45.42349 W64.34031) west onto Highway 209 and travel west for about 4.7 kilometres. Watch for a gravel pull-off beside a tall rock face on the north side of the road and immediately east of a small stream. Stop in the pull-off.

Where to Park

Parking Location: N45.42175 W64.39710

Park in the pull-off.

Walking Directions

The outcrop is beside the parking location.

Notes

The orientation of the rock layers at this site makes the outcrop prone to rock slides.

1:50,000 Map

Parrsboro 021H08

Provincial Scenic Route

Glooscap Trail

On the Outcrop

A layer of phyllite preserves small-scale drag folds that formed as the result of fault movements (see 10-centimetre scale, foreground left).

Outcrop Location: N45.42175 W64.39710

Note that the strong cleavage at this site makes the rock crumbly and is parallel to the slope of the hillside. For that reason the outcrop is unstable and not suitable for climbing. The rock is phyllite, formed by low-grade metamorphism of siltstone. The high mica content of the rock gives it a silky lustre.

In some parts of the outcrop, the cleavage is folded like a rumpled carpet. Known as drag folds (photo above), they formed during fault movement as layers below the folds moved toward the right and layers above the folds moved toward the left.

Siderite vein.

You may also see some dark grey or rusty brown veins cutting across the cleavage surface (see detail). They contain siderite, an iron carbonate mineral related to iron oxide-copper-gold mineralization along the fault system. Farther east in Londonderry, such an occurrence was the focus of Nova Scotia's largest iron mine, active from 1849 to 1902.

FYI

- The Cobequid-Chedabucto fault system is often depicted as one continuous line, as in the map at right. But it's actually a complex series of related faults, parts of which mark the boundary between Avalonia and Meguma (shaded green and blue, respectively).

Related Outcrops

Seen in the map below as an east–west line of closely spaced topographic contours, the Cobequid fault scarp forms a nearly continuous ridge along sections of Highways 2 and 209 near Parrsboro. Just west of Parrsboro a small roadside outcrop of fault-related phyllite can be seen where Kirkhill Road climbs the scarp (N45.42396 W64.36524). The scarp can also be viewed along the shore, for example, about 200 metres west (N45.40639 W64.56927) of Wards Brook Road near Port Greville.

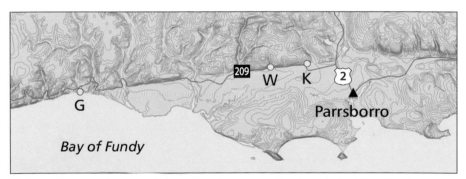

Highways 2 and 209 follow the Cobequid fault scarp near Parrsboro (G, Port Greville; W, Wharton; K, Kirkhill).

Exploring Further

For more information and photos related to the history of iron mining in Londonderry, visit Nova Scotia Archives' *Men in the Mines* at novascotia.ca/archives/virtual/meninmines/iron.asp.

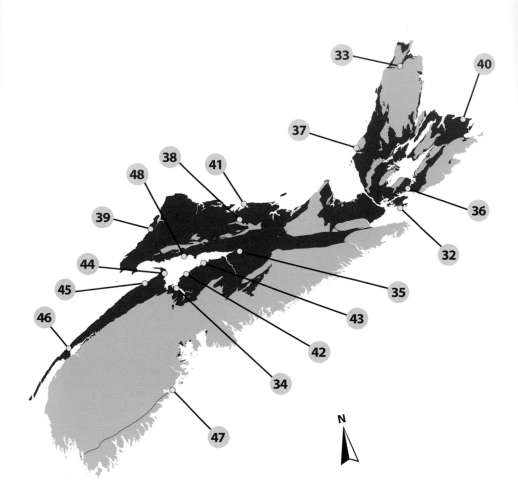

PANGAEA

At a Glance

Locations
Northumberland shore; northern mainland; Fundy shores; Cape Breton Island

Origin
Continental interior of Pangaea

Key Features
Coal, fossil forest life
Windsor Sea deposits
Red continental sediments
Thick lava flows

At these sites, you can …

32	**Marache Point**	Seek out jumbled debris from the young Appalachians.
33	**Beulach Ban Falls**	View the effects of a reactivated fault.
34	**Blue Beach**	Acquaint yourself with a lively Devonian scene.
35	**Victoria Park, Truro**	Stroll by old river deposits in a lovely river valley.
36	**St. Peters Battery**	Inspect a mineralized rock intruded along a fault.
37	**Finlay Point**	Discover what happened when a sea dried up.
38	**Balmoral Mills**	Study gravel swept from ancient highlands.
39	**Joggins Cliffs**	Wonder at a famous series of buried forests.
40	**Glace Bay**	Trace a coal seam in a coal mining town.
41	**Cape John**	Witness the start of an extreme climate shift.
42	**Rainy Cove**	Decipher the meaning of a dramatic unconformity.
43	**Burntcoat Head**	Explore red rocks typical of a desert valley.
44	**Cape Blomidon**	Gaze at the expansive remains of a desert lake.
45	**Halls Harbour**	Recognize lava flows that once smothered the land.
46	**Point Prim**	Admire the geometry of cooling basalt.
47	**Cherry Hill**	Trace a long, narrow crack in Pangaea's crust.
48	**Five Islands**	Walk beside layers from the dawn of the dinosaurs.

In Nova Scotia, rock formations deposited upon or cutting across those of Laurentia, Ganderia, Avalonia, and Meguma belong to the story of their successor, the supercontinent Pangaea. From 360 to 270 million years ago, its giant pieces moved into their final positions. Briefly, from 270 to 200 million years ago, the Earth's landmasses were unified. By 190 million years ago, Pangaea's fragmentation had begun.

Marache Point (site 32).

Pre-existing fault zones in the region were repeatedly reactivated throughout this process. Along these faults about 360 million years ago, a network of steep-sided basins formed within the young Appalachian mountains (sites 32, 33). Over time the basins broadened to host lakes or estuaries and wide river valleys (sites 34, 35). Continuing tectonic stresses (site 36) caused volcanic activity near the fault zones.

Rapid erosion eventually flattened the landscape, which then lay near the equator. From about 340 to 325 million years ago, sea level rose, then fluctuated, periodically flooding the region. Limestone, gypsum, and halite were deposited when the hot climate caused severe evaporation (site 37). After the sea withdrew for the last time, renewed uplift along the Cobequid-Chedabucto fault system led to rapid erosion, depositing thick wedges of coarse sediment nearby (site 38).

Blue Beach (site 34).

St. Peters Battery (site 36).

From about 325 to 305 million years ago, an arm of the ocean reached into the region, and the equatorial climate was tempered by distant polar ice caps. Great rivers crossed the low-lying areas, depositing sand and silt. This environment hosted lush, swampy forests, the remains of which would turn to coal (sites 39, 40).

By 290 million years ago, the climate had changed, thanks to Pangaea's ongoing consolidation. Rivers flowed only during the rainy season, and the great coal-forming forests died off (site 41). Once Pangaea was complete, the region was uplifted and eroded. Unconformities are Nova Scotia's only testament to that part of the supercontinent's history (site 42).

Shifting tectonic forces began tugging Pangaea apart around 230 million years ago, forming a broad rift valley along the Cobequid-Chedabucto fault system. Still located near the equator, the region contained many landforms typical of today's desert valleys (sites 43, 44).

As rifting continued, about 200 million years ago lava poured through deep fissures in the crust, flooding the valley with basalt (sites 45, 46, 47). Later, the province's youngest rock layers formed as rivers flowed again in the same arid valley (site 48), by then populated with many reptile species, including early dinosaurs.

Burntcoat Head (site 43).

171

An inviting lane along the south shore of Arichat harbour leads toward the Marache Point lighthouse and nearby outcrops on the headland.

Rock Collection
Colourful Conglomerate Marking a New Era

"In Loving Memory," says a plaque on the little bench beside the lighthouse. "Harry & Bernadette LeBlanc, last lighthouse keepers—1970." Their occupation has passed into the pages of history. For more than 40 years an automated light has marked the entrance to Arichat harbour. Times have changed.

Much has changed over the broad arc of geologic time, as well. At the end of the Devonian period, the landscape at this site was not gently rolling but wrenched and torn by tectonic activity. Within the new Appalachian mountain range, a ragged line of basins and intervening ridges formed along the fault zone separating Avalonia and Meguma. Into such a basin poured boulders, pebbles, and sand of every description, creating a vibrant collection that records that dramatic environment.

After you visit the colourful outcrops by the lighthouse, perhaps you'll sit on the bench if the weather is fair. Let the view of Isle Madame's fields and shores draw you into the past as you contemplate the history of the lighthouse; of Isle Madame and its proud Acadian heritage; of the restless young Appalachian mountains.

Getting There

Driving Directions

From Highway 206 in Arichat on Isle Madame, turn south (downhill) and follow any of several side streets, for example, Conney's Lane, to the waterfront. Turn east (left) and follow Veterans Memorial Drive to Robins Road (N45.51147 W60.98820) at the head of Arichat harbour. Travel west on Robins Road, turning south (left) at Cape Auguet Road (N45.50655 W61.00532). Cape Auguet Road follows the shoreline and is paved all the way to the parking location.

Where to Park

Parking Location: N45.48456 W61.02906

Park in the small gravel area on the left, at the end of the paved road. If you have a higher clearance vehicle, or if conditions are dry, you may be able to drive farther along the unpaved extension of the road. There is a larger parking area about 60 metres farther west.

Walking Directions

Follow the unpaved vehicle track about 600 metres to the lighthouse. Walk past the lighthouse to access outcrops along the shore. In wet conditions you may encounter a small stream about 400 metres from the parking location.

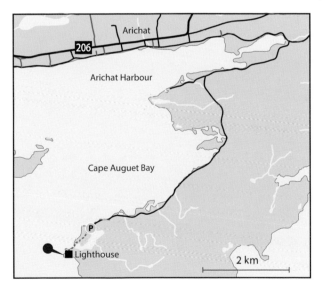

1:50,000 Map	Provincial Scenic Route
Chedabucto Bay 011F06	Fleur de Lys Trail

On the Outcrop

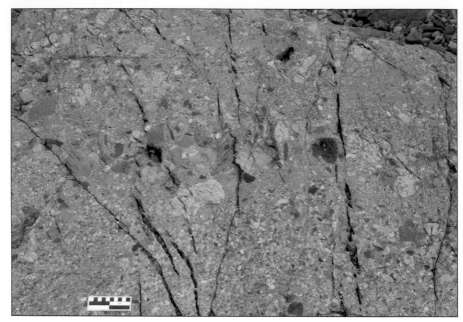

The conglomerate here contains several kinds of volcanic and sedimentary rock. In this outcrop it is poorly sorted, that is, the fragments include a wide range of sizes and shapes.

Outcrop Location: N45.48068 W61.03453

The outcrop location is on the shore west of the lighthouse. On clean wave-washed rock pavements and boulders you can see many sizes and types of rock fragment in this colourful conglomerate. For example, look for greenish basalt, dark red sandstone, light grey siltstone, and some red or pink volcanic rocks. You may even see fragments of older conglomerate.

Nearby you may find layers or irregular areas of sandstone within the conglomerate. The smaller grain size indicates water was flowing more slowly, perhaps due to a shift in the location of the main river channel.

Sedimentary layers here are tilted and become older toward the southeast. Due to the large size of many fragments, it is hard to discern the layering in some outcrops. In the sandy areas, you may see some cross-bedding due to the action of river currents.

Sandstone and conglomerate.

1000   900   800   700   600   500

Z₁   Z₂   Z₃   €

FYI

- The conglomerate at Marache Point is part of a larger rock formation at least 5 kilometres thick. The outcrops here are near the bottom of this thick formation, only about 250 metres from the base. Fragments were derived from older rock units in the adjacent highlands.

- Conglomerates like those on Isle Madame accumulated in structures called half-grabens related to movement on the Cobequid-Chedabucto fault system. A half-graben forms when a block of crust becomes tilted as one side slides downward along a fault (see diagram).

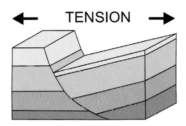

Half-graben structure.

- The wide size range and somewhat angular shapes of the fragments in the conglomerate are signs of an immature sediment carried by rapidly moving water and deposited quickly. In this case, water rushed and tumbled down the steep side of the half-graben.

Related Outcrops

The first sedimentary layers to form after Meguma joined Avalonia are known as the Horton Group (shaded red in the map at right).

The rocks at Marache Point and those at Beulach Ban Falls (site 33) are both near the base of the Horton Group. Marache Point is located along the Cobequid-Chedabucto fault system between Avalonia and Meguma, while Beulach Ban Falls lies along the Aspy fault zone in Ganderia. These widely separated faults were both reactivated at the same time due to the ongoing assembly of Pangaea.

500   400   300   200   100   0

€   O   S   D   C   P   Ŧ   J   K   Cz

Fed by Beulach Lake 350 metres above the North Aspy River, this stream cascades down tilted layers of conglomerate and sandstone at Beulach Ban Falls.

Aspy Activity
Movement along an Old Break in the Crust

You're likely to hear the *beulach* before you see the *ban*: True to its name,* the waterfall fills the surrounding woods with soothing sound, then catches your eye with a dash of white among the foliage. The cascade is one of many expressions of the Aspy fault, a prominent topographic feature of northern Cape Breton Island and one of great longevity.

Late in the Devonian period, fault movements created a steep-sided basin in this area, similar to others in the region (see site 32). Upland trees had not evolved yet, and rivers freely eroded the exposed high ground. Rock fragments of all sizes were swept into the basin, eventually creating a thick wedge of sedimentary rock. The present-day stream bed below Beulach Ban Falls, dotted with boulders and other loose stone, provides a telling analogy.

The Aspy fault of today approximately follows the ancient boundary between contrasting regions of Ganderia, a site of crustal weakness along which movement has occurred numerous times.

---

* In Scots Gaelic, *beulach* means talkative or prattling; *ban* means white or pale.

Getting There

Driving Directions

Along the Cabot Trail about 10.5 kilometres southwest of Cape North and near the Big Intervale campground, watch for signs to Beulach Ban Falls. Turn where indicated—there is separate access for north- and southbound traffic. Follow the gravel road about 2.2 kilometres to the parking area.

Where to Park

Parking Location: N46.81408 W60.62635

Park in the open area near the trail head.

Walking Directions

From the east side of the parking area, follow the trail east through the woods, along the north side of the stream. The trail ends in view of the falls.

Notes

This site lies within the boundaries of Cape Breton Highlands National Park and requires a valid park pass.

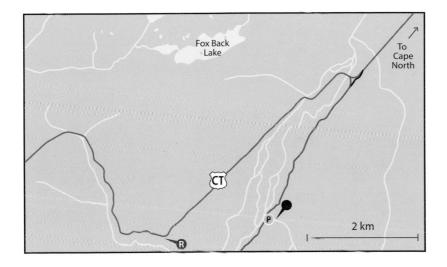

1:50,000 Map

Pleasant Bay 011K15

Provincial Scenic Route

Cabot Trail

On the Outcrop

Tilted layers of red sandstone and coarse conglomerate are clearly visible in the cliffs beside the waterfall.

At this site, tilted slabs of sandstone and conglomerate form a chute down which the waterfall flows. The cliffs beside the falls provide clear views of the rock layers. When the water level is low, conditions may allow you to examine outcrops and boulders in the riverbed, too.

Many of the layers contain rock fragments of significant size, 50 or more centimetres wide, indicating the power of the ancient river. Most are rounded, having been worn and shaped in the ancient riverbed. They represent a variety of rock types found in the Cape Breton Highlands (see sites 13–15), which were exposed (as they still are today) in the high ground above this site.

Deformed sedimentary rock.

Periodic reactivation of the Aspy fault accounts for both the conglomerate and the present waterfall. Beside the footbridge near the parking area, look for a dark grey, fine-grained sedimentary rock on the south side of the stream (see detail). Folded layers in the rock show the effects of deformation along the Aspy fault.

1000   900   800   700   600   500

Z_1 Z_2 Z_3 $\in$

By the Way

In 1883 geologist Hugh Fletcher wrote of the North Aspy River: "So remarkably straight is this river that one can obtain a lovely view down the valley to the sea from near its source, and a continuation of this straight line skirts the high hills of the promontory of Cape North."

You can share in his experience as you travel the Cabot Trail. Lookoffs (N46.87087 W60.54698 or N46.81234 W60.64251) near Beulach Ban Falls provide dramatic views of the valley and the cape cliffs beyond.

For additional views, visit Cabots Landing Provincial Park (N46.94282 W60.46286). To reach the site from the Cabot Trail, in the community of Cape North turn (N46.88463 W60.50654) onto Bay St. Lawrence Road and continue about 10 kilometres to the park entrance.

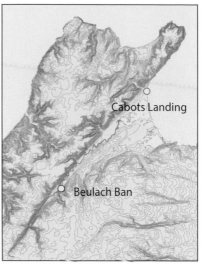

Topographic contours emphasize the Aspy fault trace along the North Aspy River (shaded bright blue) and the southeast shore of Cape North.

Cabots Landing Provincial Park provides dramatic views of the Aspy fault scarp.

Exploring Further

Fletcher, Hugh. "Report on the Geology of Northern Cape Breton." In *Report of Progress, 1882-83-84* (Section H). Dawson Bros., 1885, pp. 79–80.

Parks Canada information about the Aspy Trail, www.pc.gc.ca/eng/pn-np/ns/cbreton/activ/randonnee-hiking/aspy.aspx.

00   400   300   200   100   0

| € | O | S | D | C | P | R | J | K | Cz |

Erosion of gently tilted siltstone and shale layers provides a continual supply of rock fragments on the shoreline at Blue Beach.

Nature's Archive
Fossil-Rich Rocks of a Late Devonian Shoreline

Twice a day the Fundy tide rises and falls across the wide shore at Blue Beach. Year by year as sea level subtly rises, the low, layered outcrops along the shore gradually disintegrate. The beach is littered with fragments broken from the adjacent cliffs. Like shreds and tatters of an ancient scroll, to a practiced eye these fragments have a tale to tell.

At the end of the Devonian period, all of what is now Nova Scotia lay near the equator. This site was teeming with life and was situated in a watery landscape of brackish estuaries, tidal flats, and lowlands. You don't have to look far for an analogy: Low-lying areas where rivers meet the Fundy Shore are a good match.

The Devonian world represented here had started to resemble our own in some basic ways. The land was inhabited by plants, some the size of those in the woods now bordering Blue Beach. Recognizably modern fish swam in the sea. A stroll along the shore may bring you glimpses of somewhat familiar, though ancient, signs of life.

Getting There

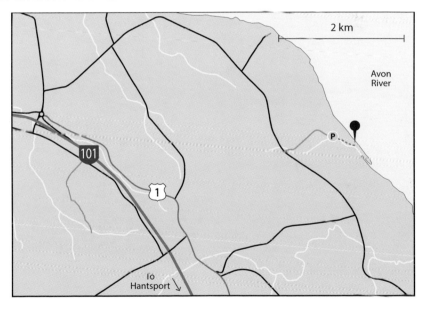

Driving Directions

From Highway 1 northwest of Hantsport, access Bluff Road—for example, directly from Highway 1 near Hantsport, or via Oak Island Road from the roundabout near Highway 101, exit 9. Follow Bluff Road to Blue Beach Road, turning (N45.09630 W64.22261) east to follow Blue Beach Road for about 625 metres.

Where to Park

Parking Location: N45.09786 W64.21590

Park in the open area at the end of Blue Beach Road.

Walking Directions

From the southeast corner of the parking area, follow the trail about 250 metres through the woods to the shore. As tide conditions allow, turn left (northwest) and explore along the beach for another 250 metres.

Notes

This site is best visited in low tide conditions.

1:50,000 Map

Wolfville 021H01

Provincial Scenic Route

Evangeline Trail

On the Outcrop

Low cliffs along the shore contain tilted, folded layers of pale brown siltstone and dark grey shale. Fragments of both are plentiful on the beach also.

Outcrop Location: N45.09716 W64.21284

Where the trail emerges onto the shore at Blue Beach, layers of pale brown siltstone and dark grey shale alternate in the low cliff. The layers preserve sedimentary features giving clues to the ancient environment. They also preserve fossils and tracks of the site's ancient inhabitants. Rock fragments on the beach display a rich variety of these features.

As you explore, remember that collection of fossils in Nova Scotia is severely restricted. No one may collect specimens from an outcrop without a permit issued by the province. Small rock fragments on the beach are not covered by this restriction, but here—as at any site—it's best to take photographs only.

At low tide on some parts of the beach, siltstone layers form long, narrow platforms extending out into the mud flats. As conditions allow, and with appropriate footwear, you may wish to follow a siltstone layer. Be sure to return, then follow the next rather than crossing the mud—it is very sticky and tends to stain clothing. As always, know the timing of tide movements for the day of your visit.

On the Outcrop (cont'd)

Sedimentary structures preserve the effects of the environment, including (**a**) ripple marks in siltstone and (**b**) mud cracks in shale.

Sedimentary Structures

Sedimentary structures are, in a sense, fossilized events. You may have noticed that even the smallest mud puddle can record the effect of a dry spell or a rain shower, for example. The same has been true throughout geologic time. Here are two common examples found on Blue Beach.

Some siltstone layers preserve ripple marks. A typical example (photo **a**) has long straight ridges with more or less symmetrical sides. This indicates the ripples were shaped by wave action rather than by flowing water. You may have seen similar ripples in the sand of a lakeshore or beach—even this beach—after a spell of breezy weather.

In some slabs of shale you may find a pattern of sharply defined, criss-crossing ridges (photo **b**). They formed in a two-stage process. First, an existing layer of mud was exposed to air and dried out, forming cracks. Then a second layer of mud or silt was deposited onto the older, cracked layer. The ridges you see formed on the bottom of the second layer, as wet sediment flowed into the underlying cracks.

On the Outcrop (cont'd)

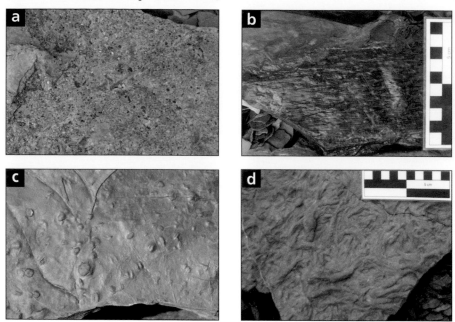

Fossilized remains include (**a**) sparkling, black fish scales and (**b**) black tree bark. Trace fossils appear on the bottom surface of sedimentary beds and record the activities of (**c**) small arthropods and (**d**) worms.

Fossils and Tracks

At Blue Beach you may find fossilized remains of plants and animals. You may also find animal tracks and trails preserved in stone; these are known as trace fossils.

Sparkling black fish scales are fairly common in the shale. In most cases, only scale fragments are preserved (photo **a**). If you find whole fish scales, look for tiny concentric ridges, or growth lines, following the scale outline. Plant remains are common here, too, especially the bark of the tree-like *Lepidodendron*. The black surface of this fossil is dimpled where foliage once grew from the trunk (photo **b**).

Animals crawling or feeding on the top surface of the sediment left all sorts of trails and dents recording their activities. The next layer of sediment filled in these marks, preserving them as bulges on the bottom of the new layer. For example, a type of small arthropod (related to modern insects and crabs) left little dents like a coffee bean (photo **c**) wherever it paused to eat. A wide variety of worms left trails in the mud as well (photo **d**).

1000		900		800		700		600		50
	Z_1				Z_2				Z_3	$\mathbb{C}$

FYI

- The rock layers at Blue Beach are intensely folded on a large scale. At low tide, especially from higher ground, you can see some of the fold patterns traced by prominent, more erosion-resistant siltstone layers.

- At the base of this rock formation (not visible here) is an unconformity, below which are rocks of the Meguma terrane: the South Mountain Batholith (see sites 29 and 30) and the Rockville Notch, Halifax, and Goldenville groups (see sites 17–25).

- Clear or colourless sparkling flecks in the rocks are grains of muscovite weathered from granite of the South Mountain Batholith.

- Fossil footprints and bones from Blue Beach and sites of similar age have been important to understanding the evolution of reptiles, which first appeared late in the Devonian period.

The trail to Blue Beach leads through the woods on the clifftop.

Related Outcrops

Slightly younger layers of siltstone and shale can be seen along Lepper Brook in Victoria Park, Truro (site 35). They formed in the flood plain of a river and contain few fossils compared to Blue Beach.

Exploring Further

For more information about restrictions on fossil collecting in Nova Scotia, visit the website of the Museum of Nova Scotia at museum.novascotia.ca/about-nsm/about-heritage/special-places-protection-act.

500   400   300   200   100   0

Є   O   S   D   C   P   ℞   J   K   Cz

Wide, well-kept trails follow Lepper Brook, which flows through a forested gorge in Victoria Park.

Valley in a Valley
Early Carboniferous Flood Plain Deposits

In the centre of Truro is more than 400 hectares of natural beauty. Step into Victoria Park and you step from the bustle of town into a world of hills, mature forest, wildlife, rocky cliffs, and waterfalls. You also step into history: The park was established in 1887, making it one of Nova Scotia's oldest protected areas.

Lepper Brook flows north and west through the park, over waterfalls and into a deep gorge. Trails along the brook lead past dramatic outcrops of steeply tilted siltstone and shale. The splashing brook provides the perfect soundtrack for viewing these sedimentary rocks, which are also of river origin. They formed during floods in a broad, flat valley early in the Carboniferous period as rapid erosion was wearing down the young Appalachian mountains.

In fact, a recurring theme emerges in the geologic history of Victoria Park. Where Lepper Brook crosses the flat ground near park amenities, evidence of another river, this one of Triassic age, can be seen in the present-day stream bed. Like some other sites in the province, its past is reflected in the present environment.

Getting There

Driving Directions

From Highway 2 on the west side of Truro, follow Prince or Arthur Street to Young Street and watch for signs to Victoria Park. Follow the signs from Young onto Brunswick Street and watch for Park Road. Turn (N45.36171 W63.27401) south onto Park Road and follow it into the parking area.

Where to Park

Parking Location: N45.35875 W63.27358

This is a large parking lot beside a grassy area with park amenities.

Walking Directions

From the parking location, follow the wide, paved trail toward Lepper Brook and cross the bridge. Turn right and follow the trail that winds along the east side of the brook. By the footbridge with a gazebo on the far side is a large rock cliff—the outcrop location.

Notes

Victoria Park has an extensive system of trails, bridges, and stairs. For more information and a map of the park, visit www.truro.ca/vic-park.html.

1:50,000 Map

Truro 011E06

Provincial Scenic Route

Glooscap Trail

187

On the Outcrop

River deposits are exposed in a dramatic cliff face, as seen here from the stairs behind the gazebo. The resistant siltstone layers are prominent; weaker shale layers are recessed. The tops of the layers are on the right.

Outcrop Location: N45.35278 W63.27213

As you view the outcrop, think about the kind of flood that makes news headlines across Canada and the world. When rivers burst from their banks, sediment-laden water rushes into adjacent low-lying areas. There it stands for days or weeks as the flood slowly recedes.

These layers were deposited in events just like the newsworthy floods of today. Each siltstone layer formed in a single episode as silt (fine sand) settled from rapidly overflowing water. Mud settled out slowly afterward, later becoming shale. Intervening, less severe floods may have deposited additional mud before the next major breach brought another layer of silt.

Deformed sand dyke in shale.

You may notice an unusual detail in this outcrop. Narrow wedges of grey siltstone cut across the adjacent shale in a few locations. These wedges are known as sand dykes. They formed after the sediment was deposited but before it hardened into stone. See FYI for more details.

1000		900		800		700		600		500
	Z_1				Z_2			Z_3		$\in$

FYI

- The sand dykes may have formed during an earthquake or other quick shock. The still-pliant mud layers cracked, and into the cracks poured grains of silt from the layer above. The resulting carrot-shaped dykes are about 5 centimetres wide at the top (or right-hand) end and about 70 to 80 centimetres long.

- As the mud and silt were buried by more sediment, compaction caused the once-straight dykes to crumple, especially at the narrow end (see photo, previous page). Mud contains a lot of water and compresses easily. Silt contains less water and resists compression. To accommodate the narrowing mud layer, the sand dyke had to crumple.

A grey sand dyke cuts across a layer of reddish shale.

Related Outcrops

In low water conditions you may be able to see much younger (Triassic) river deposits in the stream bed near the park's picnic area. By the bridge leading to the trails, a guardrail extends northward along the west side of the brook. Walk about 20 metres past the end of the guardrail (N45.35792 W63.27314).

Looking back toward the bridge (on your right), the outcrops you see underwater are the tilted, grey Carboniferous river deposits so common throughout the park. But in front of you, the water flows over flat-lying rock layers with an orange colouration—the Triassic river deposits. Between the two rock types is an unconformity representing more than 100 million years.

Exploring Further

To download the brochure, "A Walking Tour of Rocks, Minerals, and Landforms of Victoria Park, Truro," visit the Nova Scotia Department of Natural Resources online at novascotia.ca/natr/meb/pdf/ic11.asp.

00   400   300   200   100   0

Є   O   S   D   C   P   Ṛ   J   K   Cz

St. Peters Battery, originally positioned for military advantage, still provides excellent views of local scenery.

From Far Below
A Fault-Related Carboniferous Intrusion

For more than 150 million years, from late Devonian to early Permian times, layer upon layer of sediment accumulated in a region known as the Maritimes Basin, forming sedimentary rocks that now cover much of Nova Scotia and neighbouring provinces. Sedimentation was the name of the game for this region. So why is this site about igneous, not sedimentary, rock?

The Maritimes Basin formed during the final stages of Pangaea's assembly, as the small terranes between Laurentia and Gondwana were jostling and shifting. Much of the movement among them was accommodated on pre-existing fault zones. Weaknesses and tears along the fault zones created room for small igneous intrusions. Here they also provided a conduit for hot fluids rising through the crust (see FYI).

At St. Peters Battery, you can see outcrops of gabbro that crystallized 339 million years ago. The gabbro is younger than sedimentary rocks of the Horton Group (see sites 32–35) but older than those of the overlying Windsor Group (see site 37). The gabbro may have been intruded here due to the site's proximity to a major branch of the Cobequid-Chedabucto fault system.

Getting There

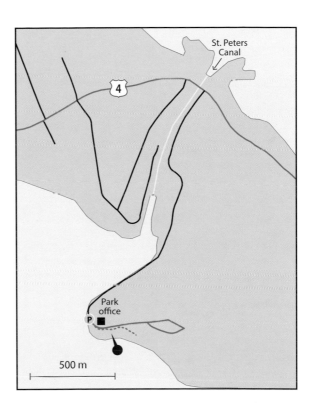

Driving Directions

Along Highway 4 east of St. Peters and immediately east of the bridge over St. Peters Canal, watch for a provincial park sign and turn (N45.65700 W60.86676) south into Battery Provincial Park. Follow the road about 1 kilometre to the small lighthouse and flagpole near the park office.

Where to Park

Parking Location: N45.64858 W60.87281

This is a small parking area near the lighthouse, flagpole, and park office.

Walking Directions

From the parking location, follow the park's gravel road around the side of the park office, then fork right onto a grassy hiking trail. Fork right again onto a narrower trail and follow it into the woods and toward the shore for about 100 metres. As conditions allow, make your way down to the beach, then turn right and double back for about 90 metres to the outcrop.

Notes

This site is located within Battery Provincial Park. There is no cost for day-use visitors.

1:50,000 Map

St. Peters 011F10

Provincial Scenic Route

Fleur de Lys Trail

On the Outcrop

The gabbro at this site was fractured in a variety of ways, resulting in features such as (**a**) gabbro breccia, (**b**) a fine-grained gabbro dyke, and (**b, c**) discolouration along mineralized cracks.

Outcrop Location: N45.64806 W60.87164

As you emerge from the woods onto the beach, you may first notice an unrelated outcrop of black shale, likely part of the Horton Group. Once on the main outcrop, you'll see signs that the St. Peters gabbro had an eventful history. Parts of the gabbro look like a pile of rubble, while some surfaces are laced with salmon-coloured cracks.

The areas that look broken up (photo **a**) formed by a process known as auto-brecciation while the intrusion was cooling just below the Earth's surface. Movement of the magma as it rose in the crust fractured parts of the intrusion that had already crystallized.

A few narrow dykes of slightly younger, fine-grained gabbro cut across the pluton (photo **b**). They later served as conduits for mineral-rich fluids, as indicated by pale zones of alteration. The same fluids travelled through other, less regular, cracks as well (photo **c**). Within the pale areas around the cracks are millimetre-wide veins filled with a bright orange mineral, siderite (iron carbonate).

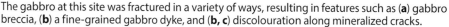

FYI

- On the outcrop, you may find small, irregular areas filled with a cloudy white mineral, calcite (see photo). The calcite was deposited by circulating fluids.

- The orange siderite here is typical of the iron oxide-copper-gold mineralization that is associated with Carboniferous intrusions along the Cobequid-Chedabucto fault system. Occurrences are found as far west as Saint John, New Brunswick, where the fault system extends through the Bay of Fundy.

- Even if crustal blocks are moving side by side, space can open along a fault with an irregular boundary, in a process known as transtension (see diagram). As happened with the St. Peters gabbro, magma may rise into such spaces, bringing heated, mineral-rich fluids like those associated with iron oxide-copper-gold occurrences.

- The Dead Sea is a present-day feature formed by transtension along the fault zone between the African and Arabian tectonic plates. Moving at an average rate of 5 millimetres per year during the last 5 million years, the fault has opened a rift extending as far as 700 metres below sea level.

Calcite deposits in gabbro.

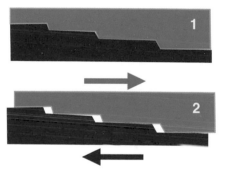

Transtension caused by transcurrent motion along an irregular fault (map view).

Related Outcrops

If you can't access the shore when you visit, you can view the gabbro from above, along the fence line near the lighthouse (N45.64821 W60.87265).

Other evidence of iron oxide-copper-gold mineralization in the Cobequid-Chedabucto fault zone can be seen along the fault scarp near Parrsboro (see site 31). However, in that location, the mineralization occurs in a deformed sedimentary rock.

White outcrops of gypsum and anhydrite are conspicuous in the shoreline cliffs at Finlay Point near Mabou Mines.

Hot and Salty

Rocks Formed from Evaporation of Sea Water

Have you seen them in the distance on your Nova Scotia travels? The ghostly white, pillared cliffs known as hoodoos can be seen in the central mainland and in many parts of Cape Breton Island. Their appearance is extraordinary, yet the gypsum they contain has more practical associations—plaster, wallboard, paint, cement—the stuff of everyday life.

Nova Scotia's gypsum is also part of a story about global climate change. For about 15 million years during the Carboniferous period, global sea level rose and fell repeatedly. Many geologists think the fluctuations were due to growth and shrinkage of an ice cap on the supercontinent Gondwana, parts of which lay near the south pole.

When sea level was high, ocean water flooded much of what is now Nova Scotia, forming a body of water known as the Windsor Sea. The whole region lay near the equator, so as sea level dropped, the Windsor Sea slowly evaporated under a tropical sun, in an environment much like parts of the modern-day Persian Gulf. The water became more and more salty, eventually precipitating a variety of minerals, including gypsum.

Getting There

Driving Directions

From Highway 19 just east of the bridge in Mabou, turn west onto Mabou Harbour Road and follow it for about 4.25 kilometres. Turn (N46.08615 W61.44376) right (north) onto Mabou Mines Road and follow it north for about 6 kilometres. Where Mabou Mines Road curves right, instead fork (N46.13251 W61.45950) left onto Mabou Mines Branch and follow it down to the harbourfront.

Where to Park

Parking Location: N46.13476 W61.46207

Park out of the way of harbour traffic on the north side of the building.

Walking Directions

From near the northwest corner of the building, follow a grassy gravel track west to the beach. As conditions allow, walk along the beach for about 150 metres to view the outcrops.

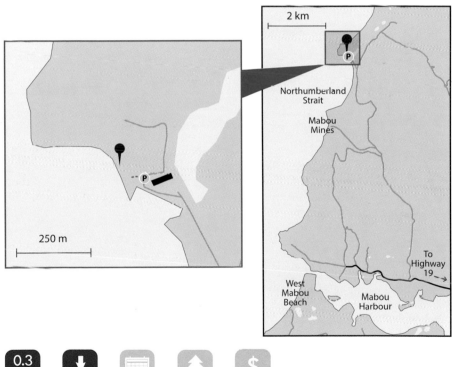

1:50,000 Map

Lake Ainslie 011K03

Provincial Scenic Route

Ceilidh Trail

On the Outcrop

Gypsum is slightly soluble in water, so it weathers easily. Outcrops typically have a crumbly or weather-beaten appearance, as seen here.

Outcrop Location: N46.13507 W61.46293

At this site you can see two closely related minerals, gypsum (main photo) and anhydrite (detail). Both are found in the distinctive pale cliffs near the harbour. Gypsum is softer (weaker) and typically lighter in colour than the harder, grey anhydrite, but both are forms of calcium sulphate. In gypsum the crystals contain molecules of water ($CaSO_4 \cdot 2H_2O$); but anhydrite is, as its name suggests, anhydrous or lacking in water ($CaSO_4$). The orange stain is due to iron oxide seeping through the cracked and crumbling outcrop.

Anhydrite.

Both anhydrite and gypsum crystallized as the Windsor Sea evaporated. Deeply buried gypsum often dehydrates to form anhydrite, then when exposed near the surface again, anhydrite may change back to gypsum by interacting with groundwater. Probably due to nearby fault movements (see By the Way), the anhydrite here is severely fractured. You can see lighter areas where the conversion to gypsum is under way.

1000	900	800	700	600	50
Z_1		Z_2		Z_3	€

By the Way

Sedimentary rock formations exposed on the beach include (**a**) maroon sandstone and shale, and (**b**) pale brown sandstone containing seams of coal.

Younger Sedimentary Rocks

If you follow the cliff northwest from the harbour toward the point, you will see two types of rock very different from the gypsum. Tens of millions of years separated the formation of the rock layers in each section of the cliff. The collection was assembled by fault movements that occurred long afterward.

The crumbly, maroon rock layers (photo **a**) next to the gypsum are loosely compacted coarse sandstone and shale. They are part of the same rock formation exposed at Balmoral Mills (site 38). The pale brown layers are sandstone (photo **b**), similar to the rock at Glace Bay (site 40). Between some of the sandstone layers are seams of coal up to 1 metre in thickness.

The coal seams formed in swampy areas with abundant plant life, and you may find plant fossils at the site. Typically they take the form of dark, dimpled coatings on the surface of a sandstone fragment (see site 34 for a similar example). In or near the coal seams, wooden beams from long-abandoned mines are sometimes exposed as the cliff recedes.

00	400	300	200	100	0
Є O S D	C	P	T̶ J	K	Cz

Related Outcrops

This outcrop of dolostone by Ingonish wharf seems out of place among the many igneous and metamorphic boulders that surround it (see site 14). It is by far the youngest rock at the site, having formed as an algal mound in the Windsor Sea.

At Ingonish wharf (N46.68855 W60.35724), an outcrop of dolostone (calcium-magnesium carbonate) is accessible in low-tide conditions. It has a lumpy appearance, and on some surfaces you can see fine, curved layering. These features were created by communities of algae living in the Windsor Sea.

Halite (rock salt) is the Windsor Sea's stealthiest deposit: Because it is so soluble, it never appears above ground in Nova Scotia's climate. However, at Malagash, you can visit the site of an historic salt mine and learn about its operations at the nearby Malagash Salt Museum.

The Malagash Salt Miners' Memorial (shown here; N45.79254 W63.32872) and nearby Malagash Salt Museum provide touching reminders of earlier times in Nova Scotia's still-profitable salt industry.

FYI

- Sea water contains calcium, sodium, sulphur, and many other elements. When evaporation occurs, minerals begin to crystallize in a predictable sequence: first calcite (calcium carbonate), followed by anhydrite and gypsum (calcium sulphate), and finally halite (sodium chloride).

- Mineral deposits of the Windsor Sea have provided significant economic benefit to Nova Scotia, bringing in $100 million or more per year. The province produces more gypsum than any other part of North America, and its salt operations deliver more than 1 million tonnes each year. Quarrying of limestone and dolostone for use in agriculture and cement-making is also a major contributor to the economy.

Related Outcrops

Gypsum and anhydrite, salt, and related rock types that formed in the Windsor Sea are all part of the Windsor Group (shaded red in the map at right). As you travel Nova Scotia, the characteristic crumbling white cliffs of gypsum and anhydrite may catch your eye in a wide variety of locations, for example:

- On the mainland, from Highway 101 just northwest of Exit 4 (N44.96619 W64.02539), seen in cliffs north of the highway.

- On Cape Breton Island, from Highway 105 or the Cabot Trail near their intersection (N46.20825 W60.59381), seen along the western shore of St. Anns South Gut.

- About 2.5 kilometres farther south on Highway 105 (N46.18982 W60.61175), seen beside a small pond west of the highway.

- Along the Cabot Trail at Quarry Road near Dingwall (N46.88420 W60.49134), seen in cliffs of a large gypsum quarry inactive since 1955.

Exploring Further

MacQuarrie, John R. *Malagash Salt*. North Cumberland Historical Society/ Tribune Press, 1994.

Malagash Salt Mine Museum, 1926 North Shore Road, Malagash (N45.79234 W63.32686). In the museum you can view documents and artifacts from the historic mine as well as a vintage National Film Board documentary that chronicles the work of salt miners underground.

On the grounds of Balmoral Grist Mill Museum are outcrops of reddish-brown conglomerate deposited during the middle of the Carboniferous period.

Emerging Land
River Deposits Signalling Renewed Uplift

In the 1800s, settlers living near Matheson Brook thought it would be a good site for a mill. After all, the brook flows down smartly from the nearby Cobequid Highlands, providing plenty of power. Several mills were established here, of which the surviving Balmoral Grist Mill is now preserved as a museum.

A trip to this scenic valley is doubly rewarding, with local and natural history both on display. The layers of conglomerate exposed here were deposited by an ancient river that, like Matheson Brook, flowed energetically out of adjacent highlands. Renewed uplift along the Cobequid-Chedabucto fault system sent rivers pouring onto the broad plains left by the final retreat of the Windsor Sea (see site 37).

When a river emerges from high ground onto a flat plain, it loses energy and must drop much of the sediment it carried. The river channel can become choked, causing the water to find another way downhill. Over time, numerous channels form, building up a wedge of sediment known as an alluvial fan. The rocks around Balmoral Grist Mill formed in such an environment.

Getting There

Driving Directions

From Highway 256 (Balmoral Road) between Highways 311 and 326, turn (N45.64634 W63.19835) southeast at Peter MacDonald Road. Follow the road for about 350 metres. Parking is on the left.

Where to Park

Parking Location: N45.64443 W63.19515

Park in the open area on the east side of the road.

Walking Directions

On the east side of the parking area, follow a short trail leading through the trees toward Matheson Brook. There, a stairway leads to a bridge across the brook. Cross the bridge and turn right, then walk across the grass to view the nearby outcrop at the base of the hillside. Alternatively, from the bridge turn left and walk around the front of the mill. On the north (downstream) side of the building, another stairway leads down to the stream bed. As conditions allow, turn right at the bottom of the stairs for access to outcrops in the stream bed.

Notes

This site is on the grounds of the Balmoral Grist Mill Museum. There is a more extensive trail system along Matheson Brook and the mill stream.

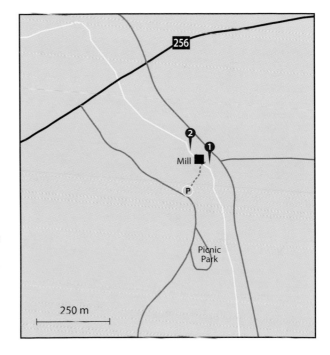

1:50,000 Map

Tatamagouche 011E11

Provincial Scenic Route

Sunrise Trail

On the Outcrop

Tilted sedimentary layering is revealed by variations in the size and distribution of pebbles and by the subtle alignment of pebble shapes in this outcrop of conglomerate.

Outcrop Location: N45.64505 W63.19447

In the vicinity of the mill are several outcrops of gently tilted conglomerate. In a crumbly matrix of reddish-brown sand and clay are varying amounts of pebble-sized, mostly angular fragments of a variety of rock types, including fine-grained red and green volcanic rock and limestone containing fossil shells.

This is a case where the presence of fossils is potentially confusing. The fossils are from the Silurian period, making them 100 million years older than the conglomerate in which they occur. The limestone containing them was eroded from elsewhere and carried here (see FYI).

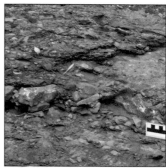

Angular fragments in conglomerate.

Pictured above is an easily accessed outcrop near the trail just southeast of the mill building. In low water conditions, you may be able to descend the stairs on the northwest side of the mill and examine river-washed outcrops. There, the jumbled variety of fragments in the conglomerate can be seen in better detail.

1000	900	800	700	600	500
Z_1		Z_2		Z_3	€

FYI

- Sediments with angular or irregular fragments are labelled by geologists as "immature." Such a condition indicates that the material did not travel far from its source, having had little chance to be worn or smoothed.

- Some geological detective work has traced the conglomerate fragments to rock formations in the nearby Cobequid Highlands. For example, Silurian limestone in the highlands contains fossils identical to those in the conglomerate. The highlands also contain volcanic rocks matching some of the conglomerate's colourful angular fragments.

Silurian limestone with worn fossil shells.

By the Way

The foundation of the mill is made of sandstone quarried in nearby Wallace. Since 1863 the Wallace quarry has supplied stone for many prominent buildings in New England and Canada, including the Parliament buildings in Ottawa. The Wallace sandstone is part of a younger group of rock layers also found at Joggins (site 39).

Related Outcrops

The conglomerate at Balmoral Mills is part of a sequence of rock layers known as the Mabou Group (shaded red in the map at right). The Mabou Group was deposited by rivers flowing across the flat region that remained after the Windsor Sea made its final retreat. Because of this, they have a distribution somewhat similar to the Windsor Group (see site 37).

500   400   300   200   100   0

€   O   S   D   C   P   R   J   K   Cz

The wide shingle beach at Joggins allows access to the famous cliffs, but Fundy tides limit hiking times. A guided tour ensures a safe, informative visit.

Forests Deep
Unique Fossil Record of a Carboniferous Wetland

Truro, Nova Scotia, July 30, 1842. My dear Marianne,—We have just returned from an expedition of three days ... I went to see a forest of fossil coal-trees—the most wonderful phenomenon perhaps that I have seen ...

So wrote pre-eminent British geologist Charles Lyell to his sister near the end of a year-long tour of geological wonders. He had travelled eastern North America from end to end—Joggins Cliffs topped his list.

For more than 170 years the site has been one of the most closely studied in the world. And no wonder—the cliffs preserve an array of 63 successive, upright forests, rooted in coal or black shale and buried in layer upon layer of sandstone. Among the giant fossil plants are the tracks and remains of forest inhabitants, including the world's oldest known reptiles.

The site has a unique ability to inspire visions of the Earth's mysterious Coal Age, as seen in displays at the adjacent Joggins Fossil Centre. In 2008 the cliffs were designated a UNESCO world heritage site. Surely Charles Lyell would approve.

Getting There

Driving Directions

From Highway 302 southwest of Amherst, turn (N45.72564 W64.25183) west onto Highway 242. Travel west on Highway 242 for about 18 kilometres and continue along the road as it becomes Main Street in the community of Joggins. Follow signs for the Joggins Fossil Centre and its parking areas.

Where to Park

Parking Location: N45.69453 W64.44930

Park in one of the designated spaces at the Joggins Fossil Centre.

Walking Directions

From inside the Joggins Fossil Centre, the south door leads to a path and stairway with interpretive panels. Descend the stairs to the beach. On the beach, turn right. The cliff face can easily be accessed for about 350 metres along the shore.

Notes

Joggins Fossil Cliffs are protected by law as a provincial Special Place, Protected Beach, and UNESCO World Heritage Site. Seasonal guided tours are available but not required for site access.

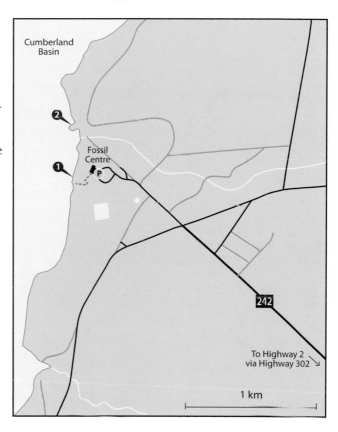

1:50,000 Map
Springhill 021H09

Provincial Scenic Route
Glooscap Trail

On the Outcrop

Continual erosion of the cliffs creates an ever-changing display of fossil plants, for example, (**a**) large *Lepidodendron*-like trunks of the main forest and (**b**) more slender *Calamites*, a form of undergrowth.

Outcrop Location: N45.69423 W64.45078

Layers at the base of the stairway lie at the top of the forest sequence. As you walk north (to the right) along the beach, you encounter older and older layers in the sequence. The rock layers include sandstone, siltstone, limestone, shale, and coal. They represent environmental conditions that fluctuated among areas of open water, marshy flood plain or delta, and well-drained lowland.

Due to Fundy tides, storm waves, and ice, the fossils visible in a given year may vary as the cliffs erode. The fossil tree trunks all tilt toward the south, since they are perpendicular to the tilted rock layers. Many of the trunks preserve surface textures, and you may find examples where the base of the trunk flares downward into fossil roots.

Some of the sandstone and siltstone layers were deposited by water flowing in river channels. You may find evidence of water flow in the form of cross-bedding (see site 41, FYI). The cliffs also contain numerous seams of coal, fragments of which often litter the beach at low tide.

1000   900   800   700   600   50

Z₁   Z₂   Z₃   €

FYI

- The fossil forests of Joggins Cliffs represent a sequence of rock layers more than 1.5 kilometres thick, and the available evidence indicates that each layer of the forest was buried over a period of just a few decades. However, the trees' upright position rules out catastrophic flooding, and the cause of their rapid burial was a perplexing mystery until recently.

  In 2005, a detailed study of underground rock structures found an explanation: Thick deposits of halite (rock salt) from the Windsor Sea once lay below them. Halite recrystallizes easily, allowing it to flow as a solid mass. Under the growing weight of overlying sand and coal, the halite was squeezed aside, causing the ground to sink rapidly. Rising water killed the upright trees, then buried them. The halite rose nearby to form salt domes.

Related Outcrops

On Cape Breton Island, beautiful exposures of river sandstone from the same time period can be seen at Cap le Moine (pull-off, N46.50306 W61.07301). For a close look at a slightly younger sequence of sandstone and coal, visit Glace Bay (site 40).

Sandstone on the shore at Cap le Moine, Cape Breton Island.

Cross-bedded sandstone, Cap le Moine.

Exploring Further

Calder, John. *The Joggins Cliffs: Coal Age Galapagos*. Province of Nova Scotia, 2012.

Joggins Fossil Centre, 30 Main Street, Joggins (N45.69453 W64.44930). For more information visit the centre's website at jogginsfossilcliffs.net.

The low cliff by the shore near the hospital in Glace Bay contains layers of sandstone and coal typical of the Sydney coalfield.

Black Gold
A Late Carboniferous Coalfield

Donkin, Glace Bay, New Waterford, Point Aconi, Sydney Mines … all mining towns atop a single great reserve, the Sydney coalfield. In 1720 a mine in Port Morien launched Nova Scotia's coal industry, shaping the destiny of communities in many areas of the province.

Late in the Carboniferous period, this site was part of an extensive delta or coastal plain. Vegetation thrived in tropical wetlands between large river channels. For hundreds of thousands of years at a time, thick layers of peat accumulated over wide expanses of the landscape. When subtle changes in sea level caused a shift in river flow, the peat was buried in sediment, beginning its transformation into coal.

Sea level fluctuated several times during this period as, seam by seam, the black gold was locked away underground. "In the cages then we drop till there's nowhere else to fall / And we leave the world behind us down the coal town road," wrote Glace Bay's native songster Allister MacGillivray. Mining is an underground occupation, but here on the shore in Glace Bay you can view a typical coal seam in broad daylight.

Getting There

Driving Directions

From the intersection of Highways 28 and 4 (Main, Union, and Commercial Streets; N46.19686 W59.95693) in the town centre of Glace Bay, follow Commercial Street southeastward across the bridge. Continue as it becomes South Street, travelling past the regional health care facility to the shore.

Where to Park

Parking Location: N46.18277 W59.93667

Park in the gravel area at the end of the street.

Walking Directions

Facing the shore at the parking location, turn left and follow a gravel track along the shore and around a small headland. As conditions allow, cross onto the beach and walk left to the base of a low bluff.

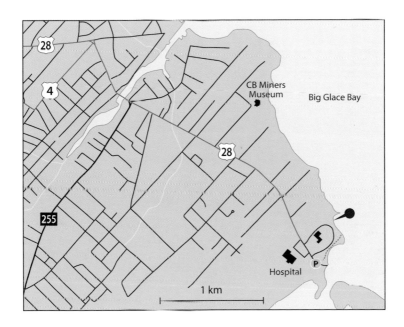

1:50,000 Map

Glace Bay 011J04

Provincial Scenic Route

Marconi Trail

On the Outcrop

This coal seam is about 60 centimetres thick; like most seams in the region, it extends for great distances. Due to its gentle tilt, this seam extends under the waters of Glace Bay, where it has been mined.

Outcrop Location: N46.18550 W59.93506

Below the coal in this outcrop is a crumbly pale rock, a layer of ancient soil. Sometimes called "seat earth" or "root clay" by miners, it commonly occurs below a coal seam. Just like the plants of today, the ancient plants from which the coal formed needed soil for physical support and as a source of nutrients.

Although the coal seam formed from plant remains, don't expect to find fossils here: Heat and pressure erase most biological structures as coal forms. In one part of the outcrop, you may notice that the coal seam has been broken and thrust over itself toward the left along a small fault (not shown above). This happened after the coal formed.

Above the coal are orange layers of river-deposited siltstone. For the best view, turn toward the northeast so you are facing the cliff at the end of the cove (with the coal seam at its base). Cross-bedding (see site 19, FYI) is one indication that they are river deposits. As well, some layers that formed in river channels have gently arched rather than parallel boundaries.

1000   900   800   700   600   50

Z₁   Z₂   Z₃   €

FYI

- The sequence of coal-bearing sedimentary rocks in this part of Cape Breton Island is about 500 metres thick. The layers formed in a series of 12 rhythmic cycles of changing climate and sea level, probably linked to the advance and retreat of great ice sheets in distant south polar regions of Pangaea.

- The rock formations seen on land extend offshore to the east for 300 kilometres or more. Petroleum exploration wells drilled east of Placentia Bay, Newfoundland and Labrador, have intersected very similar sequences of sandstone and coal. In fact, Morocco has a sequence so similar some geologists have suggested that the coalfield was split by the opening of the Atlantic Ocean.

Related Outcrops

The coalfields around Sydney are part of the Cumberland Group (shaded red in the map at right), which contains all the economically significant coal deposits in Nova Scotia.

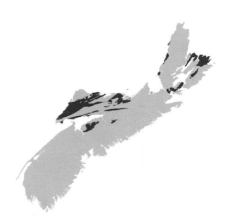

Sandstone and coal sequences at Finlay Point (site 37) and Joggins (site 39) are part of this group.

Exploring Further

Cape Breton Miners' Museum, 17 Museum Street, Glace Bay (N46.19284 W59.94231), www.minersmuseum.com. The museum is located on a former mine site and offers underground tours.

Sydney Mines Fossil Centre, 159 Legatto Street, Sydney Mines (N46.24445 W60.23487), sydneyminesheritage.ca. This part of Cape Breton Island is unique in the world for the number and variety of well-preserved plant fossils. The centre has a remarkable collection.

For the booklet *One of the Greatest Treasures* from the Nova Scotia Department of Natural Resources outlining the geology and history of coal mining in Nova Scotia, visit novascotia.ca/natr/meb/pdf/ic25.asp.

500	400	300	200	100	0
Є	O	S	D	C	P	Ŧ	J	K	Cz

A community park at Cape John overlooks shoreline outcrops of sandstone and mudstone formed during the Permian period.

Drying Up
River Deposits during Extreme Climate Change

Cape John is a unique location, one of just a few places in Nova Scotia where you can see rocks of the Permian period. It was a time of great drama in Earth history, an environmental disaster that caused massive extinctions (see site 42). The rocks here formed early in the period, as changes first took hold.

The assembly of Pangaea was driving the global climate toward hotter and drier conditions. The ice sheets over south polar regions, which had moderated the tropical climate for millions of years, collapsed and began their final retreat. Plate tectonic movements completely isolated this site and its surroundings from ocean influences. Once dominated by water-rich environments (see sites 37–40), it became a parched landscape in which rivers flowed only during seasonal rains.

Have you been in a drought-stricken area or arid region? Riverbeds are often dry, with isolated waterholes. During the wet season, run-off water flows quickly and may carry large amounts of sediment, since sparse or stressed vegetation does little to prevent erosion. The rocks here formed in such a landscape.

Getting There

Driving Directions

From Highway 6 east of Brule, about 850 metres northeast of the bridge over River John, turn (N45.75647 W63.04972) northwest onto Cape John Road. Travel about 8 kilometres—the road ends at the harbour on Cape John, near a small picnic park by the shore.

Where to Park

Parking Location: N45.79859 W63.12502

Park in the open gravel area by the harbour seawall, being mindful not to block harbour traffic.

Walking Directions

From the parking area, walk back to the picnic park. As conditions allow, make your way down onto the shore and walk right (north) about 25 metres to a rocky prominence.

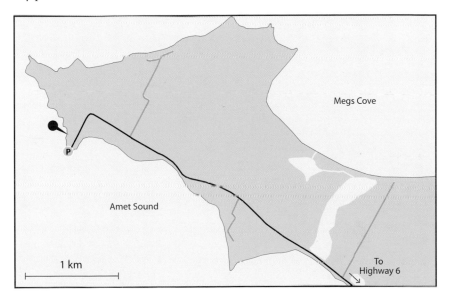

1:50,000 Map

Malagash 011E14

Provincial Scenic Route

Sunrise Trail

On the Outcrop

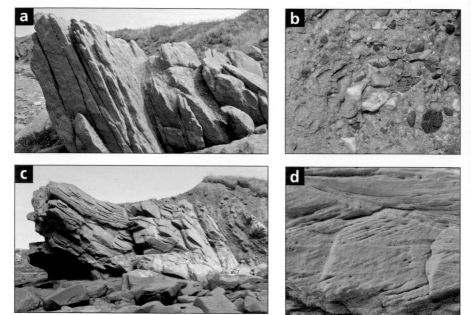

Cape John's steeply tilted sandstone layers preserve many interesting features: (**a**) coarse grey sandstone with pebbly cross-beds and (**b**) pods of gravel; (**c**) finer-grained red sandstone with (**d**) multiple sets of well-defined cross-beds.

Outcrop Location: N45.79936 W63.12508

In the shoreline cliffs, several areas of grey and red sandstone jut out toward the water. Between them are receding areas of mudstone less resistant to erosion. The waypoint marks a prominent outcrop of grey sandstone near the park (photo **a**). As conditions allow, you may be able to continue along the beach to the next sandstone outcrop, which is mainly red (photo **c**).

The sandstone preserves clear signs of its river origins. Later deformation has fractured the rock and tilted it steeply, so the first challenge is to recognize the sedimentary layering. Changes in grain size and colour are a good indication.

In the grey sandstone look for coarse-grained sediment, which indicates quickly flowing water in the main river channel, perhaps during a flood. Some areas of the outcrop are very gritty; you may also see pockets or pods of coarse gravel (photo **b**). In the red sandstone are surfaces that look as though they have been swept with gently arching strokes (photo **d**). This is cross-bedding formed as river channels meandered within a broad valley in which they flowed.

1000   900   800   700   600   50

Z_1 Z_2 Z_3 €

FYI

- These brightly coloured sedimentary rocks are known as "red beds." They are the first in Nova Scotia to record the hot, dry conditions of Pangaea's interior. The sediment grains are not red throughout, but are coated in a thin layer of iron oxide. See Geology Basics (Rock Types) for details.

- Early in the Permian period, a few isolated lowland areas were moist, allowing the growth of heat- and drought-tolerant, cypress-like conifers of the genus *Walchia*. One such area, discovered in outcrops near Brule, preserved a remarkable record of Permian forest life (see Exploring Further).

Related Outcrops

Facing Prince Edward Island across the Northumberland Strait are Nova Scotia's only outcrops of Permian-age rocks. They occur in the uppermost part of a rock unit known as the Pictou Group (shaded red in the map at right). Most of the red-brown rock layers so characteristic of Prince Edward Island also belong to the Pictou Group and are mainly Permian in age.

Permian red beds on Prince Edward Island.

Exploring Further

Creamery Square Heritage Centre, 225 Main Street, Tatamagouche (N45.71037 W63.28558), www.creamerysquare.ca. Heritage Centre displays illustrate a unique fossil discovery from nearby Brule—the only known *Walchia* fossil forest and the tracks of numerous Permian reptiles.

At Rainy Cove, gently tilted Triassic sandstone caps a tall cliff of steeply tilted Devonian siltstone and shale, marking an angular unconformity.

What's Missing?

Unconformity Marking a Crucial Earth Transition

The Permian and Triassic periods were eventful for planet Earth. Pangaea, long in the making, was finally assembled. A huge thickness of continental material was eroded away during uplift of Pangaea's interior. The climate, once tempered by ice caps, became blazing hot. At the end of the Permian period, environmental change killed off 90 per cent of all marine species and 70 per cent of land animals in the greatest mass extinction on record.

Like a traveller in crisis who stops phoning home, Nova Scotia's rock record is silent on most of these drastic events, with a gap of about 130 million years extending from early Permian to late Triassic times. Along the shore at Rainy Cove you'll see highly deformed, older rock layers directly below a wedge of younger sandstone. Between the two lies an angular unconformity, the surface that represents the missing span of time.

The site is so visually dramatic, it seems a fitting monument to the supercontinental tumult it brackets. And it holds more surprises in store. The sandstone signals another change of circumstance—Pangaea was already starting to break apart.

Getting There

Driving Directions

Along Highway 215 about 5 kilometres west of the bridge at Walton, the road runs close to the shore and crosses a small stream. Just east of the stream, watch for an unpaved lane and turn (N45.22037 W64.06761) north. Follow the lane for about 45 metres to the parking location.

Where to Park

Parking Location: N45.22070 W64.06793

Park in the small gravel area on the left (south) side of the lane, being careful not to block driveway access.

Walking Directions

Follow the lane along the shore and, as conditions allow, cross onto the beach. Follow the shoreline for about 500 metres in all. At about 250 metres, after you round a small headland, the unconformity comes into view. Continue north to the outcrop near a low, triangular opening at the base of a cliff.

Notes

This shore is subject to high Fundy tides. Access to and from the site requires low tide conditions. Due to coastal erosion, shoreline areas near the cliffs are subject to rock falls. Outcrop features should be viewed from a safe distance.

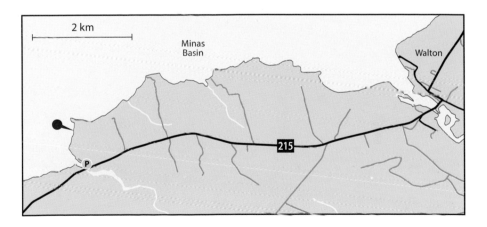

1:50,000 Map

Wolfville 021H01

Provincial Scenic Route

Glooscap Trail

On the Outcrop

Near the north end of the cliff, the unconformity intersects the shore. You can readily see the contact between contorted grey-brown siltstone and shale (below) and orange-red sandstone (above).

Outcrop Location: N45.22505 W64.07031

Note the cliff here is subject to rock falls. The site's significant features can, and should, be viewed from a safe distance.

Below the unconformity are grey and brown layers of sedimentary rock belonging to the 360-million-year-old Horton Group (see By the Way). Look for evidence of folding, traced clearly by some of the pale grey-brown siltstone layers. Even without knowing any details, their complicated, fractured appearance hints at a long history.

The rock above the unconformity is completely different. Here you find thick, intact layers of orange-red river deposits—originally sand and gravel—that are only about 230 million years old. They have been slightly tilted but are otherwise pristine. Notice that the bright colour is interrupted by irregular patches of light grey-green. These patches represent places that were starved of oxygen, perhaps due to decaying organic material, after the rock was buried.

Between the two rock types is an ancient land surface. It had ups and downs like any ordinary landscape of today. Along the unconformity are areas where the very first layer of sandstone widens or thins to accommodate the ancient surface.

FYI

- Where the unconformity intersects the beach, part of the cliff is slightly undercut at its base. Nearby, you may see a low outcrop of conglomerate that contains gravel-size pebbles as well as broken blocks of the older rock. Conglomerate is common in the Triassic rocks right above the unconformity. The pebbles gathered in low spots on the land surface as rivers began to deposit sediment across the region.

Conglomerate on the unconformity, Rainy Cove. Note that the grey Carboniferous rock occurs both as blocks within the conglomerate and as layers below the unconformity.

- Based on their thickness elsewhere, it is likely that as much as 4 kilometres of Carboniferous and Permian sediment lay above the ancient land surface here. It was removed by erosion during a period of uplift affecting central regions of Pangaea.
- After this long interval of uplift and erosion, tectonic forces began to pull Pangaea apart. The Triassic rocks of Nova Scotia are restricted to a basin that formed along the old, but reactivated, Cobequid-Chedabucto fault system in response to the new tectonic regime (see site 43, FYI).

Related Outcrops

A similar, but poorly exposed unconformity is preserved in Lepper Brook, Victoria Park, Truro (site 35). Further examples of Nova Scotia's Triassic rock formations can be seen at Burntcoat Head and Cape Blomidon (sites 43, 44).

500   400   300   200   100   0

€ | O | S | D | C | P | T | J | K | Cz

By the Way

Siltstone and shale of the Horton Group preserve (**a**) mud cracks, (**b**) tree trunks, and (**c**) animal tracks.

Carboniferous Environment

As you walk from the parking area and back, you pass by numerous folded and tilted layers of siltstone and shale below the unconformity. They are part of the Horton Group, like the rocks at Blue Beach (site 34). They formed about 360 million years ago in an area of brackish, shallow water.

At times the water withdrew, leaving the fine-grained, muddy sediment to dry out. You may see evidence of this in the form of rock surfaces that preserve mud cracks. These give the rocks the appearance of an irregular jigsaw puzzle of nearly round yet angular pieces (photo **a**).

The region lay near the equator at the time and provided a hospitable habitat for many species of plants and animals. Erosion of the cliff face creates an ever-changing assortment of examples. The site is well-known for fossil tree stumps, for example (photo **b**). On the surfaces of some layers you may find animal tracks. Those shown above (photo **c**), seen in 2013, are the tracks of *Arthropleura*, a giant millipede that became extinct during the Permian period.

By the Way

Intense folding of the Carboniferous Horton Group is evident in several locations along the shore at Rainy Cove.

Complex Deformation

As you walk to and from the unconformity at Rainy Cove, you will pass sections of the cliff face that provide dramatic evidence of tectonic activity. During the slow-moving assembly of Pangaea, this region experienced a long history of tectonic stresses, many of them expressed as episodic movements along the Cobequid-Chedabucto fault system (see site 31) between the Meguma terrane and Avalonia.

The late Devonian to early Carboniferous rock layers at Rainy Cove were deformed about 320 million years ago as a result of such movement. At a bend in the fault, rocks sitting on the Meguma terrane were compressed against Avalonia, causing them to be folded in complex patterns like wrinkles in a floor mat pushed sideways against a wall.

Later, during the Permian and Triassic periods, the area was uplifted, eroded, and exposed to the elements, much as it is again today.

Exploring Further

At a community museum in the nearby Walton lighthouse, you can view a collection of *Arthropleura* fossil tracks. For more information about the facility, visit www.novascotia.com/see-do/attractions/walton-lighthouse/1637.

At Burntcoat Head on the Bay of Fundy, Triassic river deposits frame spectacular views, ever-changing due to the site's 15-metre tidal range.

Stormy Desert
River Deposits of the Triassic Fundy Basin

From Burntcoat Head you can look across the Bay of Fundy on a clear day and see the opposite shore. The low hills on the horizon follow the trace of an ancient fault system (see site 31). During the Triassic period, as tectonic forces pulled at Pangaea's heartland, a rift formed along the fault system and it became a line of looming, rocky cliffs that shed sediment onto the valley floor below.

As you look across the bay, picture in its place a wide, asymmetrical rift valley—on the far side, the precipitous cliffs of the fault; on this side, a gentler slope upward to the south. The valley was hot and dry, swept by subtropical winds. Rivers flowed only after infrequent storms in the adjacent highlands: Water seeped into the dry ground or evaporated.

That desolate scene is a far cry from the green woods, meadows, and amenities around Burntcoat Head lighthouse today. Along the shore you can see Triassic sandstone that formed as flood waters rushed out of the highlands and spread sediment in broad sheets across the arid landscape of long ago.

Getting There

Driving Directions

From Highway 215 about 4 kilometres west of Noel, turn (N45.28567 W63.79904) north onto Burntcoat Road. Follow the road north and northeast for about 3 kilometres, watching for signs to Burntcoat Head Park. At the park access road, turn (N45.30776 W63.80177) north and drive about 500 metres to the parking area.

Where to Park

Parking Location: N45.31062 W63.80588

Park in the gravel area at the end of the park access road.

Walking Directions

From the parking area, follow a trail to the lighthouse museum, where you can sign the register and check on the day's tides if the museum is open. From the museum, follow signs and trails to a stairway among the trees. As conditions allow, descend the stairs onto the outcrop.

Notes

This shore is subject to high Fundy tides and full access to the site requires low tide conditions. Burntcoat Head Park is maintained as a local community initiative.

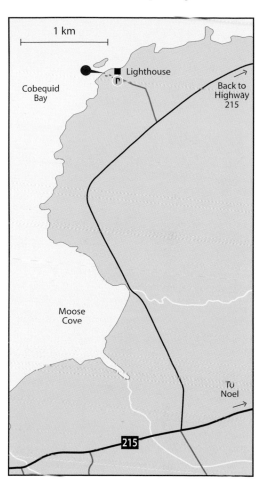

1:50,000 Map

Bass River 011E05

Provincial Scenic Route

Glooscap Trail

223

On the Outcrop

Near-horizontal layers of sandstone, each 50–100 centimetres thick, can be distinguished by contrasts in colour and texture.

Outcrop Location: N45.31142 W63.80722

Note that it is safer to explore toward the right, where a rising tide cuts off the stairway less abruptly. The rocks at this site are mostly sandstone and conglomerate deposited by rivers. Low-angle cross-bedding is common. Each layer represents a flooding event. Look for the subtle differences in colour and texture that distinguish one layer from another.

The rock has never been very deeply buried and is not strongly cemented. If you rub the surface of a rock, grains of sand may loosen under your touch. In some layers, the grains are quite round and very regular in size; such grains may have been recycled from dunes that shifted across the arid landscape.

In some layers you can see long, narrow zones of light grey rock (see detail). These sediments were starved of oxygen after burial, causing grey ferrous oxide to form instead of the red ferric oxide that colours the rest of the rock.

Grey zones in red sandstone.

1000	900	800	700	600	500
Z_1		Z_2		Z_3	€

FYI

- The rivers of this Triassic basin were not confined to individual channels but instead were braided, with many criss-crossing, shifting channels. Rivers typically take a braided form when they carry a heavy load of sediment, as they did here during infrequent, heavy rains.

- The Fundy Basin is the largest of a dozen similar basins distributed along North America's eastern seaboard. Together they represent an early, but failed, phase of ocean opening. Finally during the Jurassic period, the present-day Atlantic opened along a series of rifts farther east, now marked by the offshore continental margin.

- The Triassic basins contain evidence of mammal-like reptiles, as well as early dinosaur bones and footprints.

 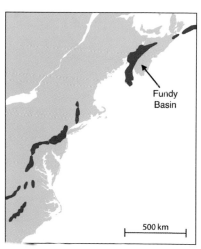

Failed Triassic rift basins of eastern North America.

Related Outcrops

The rocks here are part of the Fundy Group, which includes both sedimentary and volcanic rock formations (shaded dark and light red respectively in the map at right above). See sites 42–48 for more examples of this group.

Exploring Further

U.S. Geological Survey's "Geology of the National Parks—Death Valley Alluvial Fans," geomaps.wr.usgs.gov/parks/deva/rfan.html. Aerial photos show Death Valley landforms similar to those of the Triassic Fundy Basin, including alluvial fans and braided rivers.

500   400   300   200   100   0

€   O   S   D   C   P   Ŧ   J   K   Cz

Along the shore at Cape Blomidon Provincial Park, cliffs of orange-red sandstone and mudstone record the history of a broad valley's playa lakes.

Valley Floor
Playa Lake Deposits of the Triassic Fundy Basin

If you were to stand on the shore here at low tide on the sunniest day of summer, the expanse of Fundy mud might help you to imagine this site in late Triassic times: Think wide. Think flat. Think hot.

The rocks of Cape Blomidon were deposited in an arid, landlocked rift valley, the Triassic Fundy Basin. Earlier in its history, the steep sides of the valley had sent energetic rivers full of gritty sand pouring out of the highlands (see site 43). As erosion created a more gentle topography, only fine-grained sediment reached the valley. Floods created temporary lakes in which the sediment slowly settled.

Many terms are used to describe an evaporating lake that forms in a closed, arid valley: playa, from the Spanish word for beach; ephemeral lake, because it is temporary; salt pan or alkali flat, in reference to the salt deposited if the lake completely dries up. The full sequence of layers exposed at Blomidon represent about 20 million years of such an environment, meaning each metre of sediment in the cliffs took about 25,000 years to accumulate.

Getting There

Driving Directions

In the community of Canning from the intersection of Highways 221 and 358 (Sheffield and Main streets), follow Highway 221 east for about 1.2 kilometres and watch for Pereau Road on the left. Turn (N45.15655 W64.41480) left (northeast) and follow Pereau Road northward about 14 kilometres, passing through Pereau and Blomidon on the way to the parking location.

Where to Park

Parking Location: N45.25689 W64.35112

Park in the large open area by the entrance sign for Blomidon Provincial Park.

Walking Directions

From the parking area, follow the trail down a gentle grassy slope toward the water. Continue as the trail curves right then down a staircase through the trees to an open picnic area. Look toward the water (left), find the top of a second stairway, and descend to the beach as conditions allow.

Notes

This site lies within the boundaries of Blomidon Provincial Park. The shore is subject to high Fundy tides and site access requires low water conditions.

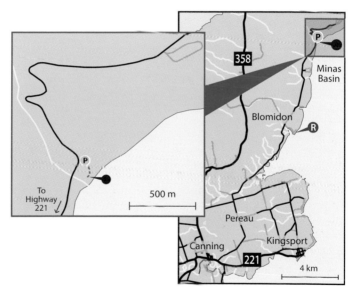

1:50,000 Map

Parrsboro 021H08

Provincial Scenic Route

Evangeline Trail

On the Outcrop

Extensive layers of consistent thickness form cycles of sandstone and mudstone in the cliff face. They are interrupted by narrow zones of grey-green rock.

Outcrop Location: N45.25582 W64.35058

The sedimentary layers exposed in the cliff face come in pairs: a thick layer of orange-red sandstone followed by a thin layer of orange-red mudstone. Typically the sandstone layers are about 1 metre thick, while the mudstone layers are about 10 to 20 centimetres thick.

Many of the original small-scale features in these rocks have been obliterated by the later growth of gypsum crystals among the sand and mud grains. This happened during dry periods when salty groundwater circulated through the sediment. Within and between some of the layers are irregular bands of grey about 1 centimetre wide. These bands became oxygen-poor after the sediment was buried, causing grey ferrous oxide to replace the orange-red ferric oxide that colours the rocks.

The regular alternation of sandstone and mudstone persists throughout the whole shoreline exposure of the cliff. In fact, 100 such cycles have been identified here. They represent shifts in the boundaries of the lake, probably due to fluctuations in climate. Sandy layers were deposited by sheets of flood water travelling toward the lake; muddy layers were deposited within the lake itself.

1000	900	800	700	600	500
Z_1		Z_2		Z_3	$\in$

FYI

- By studying the distribution of rock types in the Triassic Fundy Basin, geologists have pieced together a picture of the entire landscape. The asymmetrical valley was bordered by the steep Cobequid-Chedabucto fault scarp on the north (see site 31) and a ramp-like slope to granite highlands in the south (see site 29). It included a wide variety of landforms, including alluvial fans (A), braided rivers (B), sand dunes (D), and playa lakes (P).

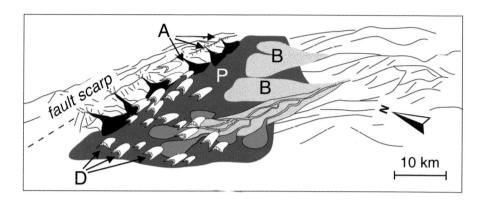

Related Outcrops

Triassic lake deposits of the Fundy Group can also be seen at Houston Beach, south of Cape Blomidon. From Pereau Road (see Getting There), turn (N45.21876 W64.37162) and follow Stewart Mountain Road southeast. Park by the shore (N45.21460 W64.36606). On the beach, as conditions allow, walk left for about 300 metres to the outcrop (N45.21669 W64.36359).

Red, gravelly clay falling from much younger glacial deposits along the top of the cliff often partially obscures the outcrop, but red and grey layers of Triassic mudstone become more obvious as you walk northward. On close inspection at this site you may see small voids in the rock left where salt crystals from the playa later dissolved.

Exploring Further

From NASA's online catalog of images, Visible Earth, visibleearth.nasa.gov/view.php?id=37536.

At this website a pair of images shows the evaporation of a playa lake in California's Death Valley. Several landforms in the image are similar to those in the Triassic Fundy Basin, with alluvial fans, braided streams, and the playa itself.

500   400   300   200   100   0

Є O S D C P Ṟ J K Cz

Layers of basalt visible in the cliffs at Halls Harbour were formed by hot lava flowing across the ground.

Rivers of Stone

Lava Flows from Pangaea's Fractured Interior

An outing to Halls Harbour is rewarding. Besides enjoying a meal or simply admiring a panoramic view of the Bay of Fundy, you can also explore Triassic rock formations along the shoreline. The rocks provide beautiful scenery now, but when they formed, a hothouse climate gripped Pangaea, leading to widespread desert conditions. This site lay in a parched rift valley far from the ocean.

At the end of the Triassic period, tectonic forces pulling Pangaea apart created a vast network of cracks from which lava emerged and spread out over parts of what is now South America, Africa, Europe, and North America—including Nova Scotia. Remnants of this aptly named flood basalt can be seen along the Fundy south shore for more than 200 kilometres, from Cape Split to Brier Island.

Lava covered the region in three major episodes totalling as much as 500 metres in thickness. At Halls Harbour, you can view multiple lava flows from the second episode. In the cliff are signs that the core of one flow once bubbled and frothed with volcanic gases, and that the bottom of the overlying flow chilled as it advanced across the ground.

Getting There

Driving Directions

In Centreville from the intersection of Highways 221 and 359, follow Highway 359 northwest about 10 kilometres toward the Bay of Fundy. The highway ends in Halls Harbour. Continue to follow the main road, which winds over a bridge and then back toward the shore, passing a complex of buildings on the right. Fork right into the parking area for these businesses.

Where to Park

Parking Location:
N45.20075 W64.62084

Park in the open area west of the buildings.

Walking Directions

From the southwest corner of the parking area, a footpath leads away from the buildings down onto the beach. Continue southwest along the beach for about 80 metres, or farther if you like—the outcrop is extensive.

Notes

This shore is subject to high Fundy tides. Site access is best in low tide conditions.

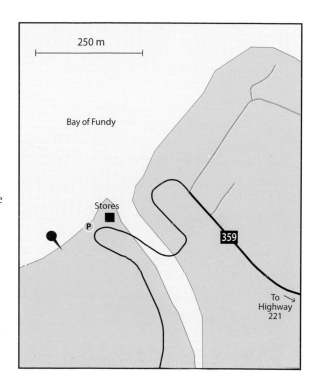

1:50,000 Map

Berwick 021H02

Provincial Scenic Route

Evangeline Trail

On the Outcrop

The chilled, uniform base of one lava flow rests on the weathered top surface of an older flow in the cliffs along the shore.

Outcrop Location: N45.20035 W64.62145

It can be helpful to walk beside the rock cliffs for about 100 metres, looking at the overall appearance. In the cliff face, two separate lava flows are visible. The upper one spread across the top surface of the lower, older one.

Before the upper flow arrived, the surface of the older flow became weathered, stained, and weakened by the harsh desert environment. Along the contact between the flows, you can see the weathered rock below and the solid-looking base of the younger flow above. The younger lava cooled quickly against the ground and for that reason is very fine grained.

Mineral-filled amygdales.

Parts of the lower flow have a dotted appearance because it bubbled with volcanic gases that left voids as the lava cooled. Later, circulating fluids deposited quartz and a variety of zeolite minerals, including stilbite, in the voids, creating the colourful amygdaloidal texture. Some zeolites have reacted with sea water, acquiring a bright blue-green coating.

1000	900	800	700	600	500
Z_1		Z_2		Z_3	$\mathcal{C}$

FYI

- Unlike the explosive volcanic rocks found in some other parts of Nova Scotia (for example, sites 7–9, 25), the rock here formed from red-hot molten lava that flowed from cracks in the ground. Similar eruptions take place in present-day Iceland.

- One of the most common zeolite minerals in the basalt is stilbite, which can be white, pink, or orange. Stilbite is Nova Scotia's official provincial mineral.

Related Outcrops

The North Mountain basalt occupies a long, narrow sliver along the southern shore of the Bay of Fundy (see site 47, Related Outcrops).

The North Mountain lookoff (N45.19976 W64.40758) on Highway 358 north of Canning provides a view that helps illustrate the scale of the lava outpourings. From the Annapolis Valley, the highway climbs onto a steep escarpment that cuts across only part of the thickness of the now-tilted layers. Even so, from the high lookoff, it's a long way down to the Triassic sandstone formations (see sites 42–44) onto which the basalt flowed.

View from North Mountain lookoff.

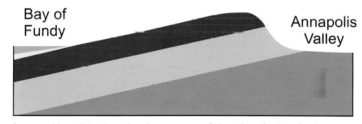

Layers of basalt (red) and sedimentary rock (grey) under North Mountain.

Exploring Further

For more information and photos of very fluid lava flows of the type seen at Halls Harbour, visit the U.S. Geological Survey's Volcano Hazards Program Glossary, volcanoes.usgs.gov/images/pglossary/pahoehoe_toe.php.

A park at Point Prim lighthouse near Digby provides access to columnar basalt exposed along the shore.

So Orderly

Basalt Columns Formed as Triassic Lava Cooled

The lighthouse at Point Prim is simple and striking in its form: a tall, four-sided column painted with handsome vertical red stripes. What a coincidence that the lighthouse and its surrounding park sit upon thousands of other, smaller columns, not quite so regular in form and mostly hidden underfoot.

At the end of the Triassic period, deep faults opened in this part of the Pangaean supercontinent. From them, red-hot lava poured out into a rift valley known as the Fundy Basin. The rock at Point Prim is basalt, formed as a final, vast sheet of lava covered the region to a depth of 150 metres.

Cooled by the ground below and the air above, the sheet of lava—initially about 1200°C—slowly lost its heat. Most solids shrink slightly as they cool, and rock is no exception. The slowly dropping temperature and even texture of the basalt caused cracks to form in a regular, roughly hexagonal pattern. The cracks spread vertically through the rock as the temperature dropped, eventually forming columnar basalt, a type of rock formation beautifully preserved here.

Getting There

Driving Directions

In Digby, from Highway 303 (Victoria Street) about 800 metres north of the intersection with Highway 217, turn (N44.63025 W65.76673) west onto Culloden (a.k.a. Raquette) Road and immediately watch for Lighthouse Road on the right. You may see signs for the lighthouse along the route. Turn (N44.63046 W65.76780) onto Lighthouse Road and follow it for about 7 kilometres to the parking location.

Where to Park

Parking Location: N44.69077 W65.78537

Park in the gravel area where Lighthouse Road ends at the metal gate.

Walking Directions

From the parking location, a footpath leads around the left side of the metal gate. Follow it, forking right away from the lighthouse to a grassy area with interpretive panels. Farther to the right, walk across a picnic area onto the rock pavement.

Notes

The lighthouse, surrounding property, and amenities at this site are maintained through the efforts of the Friends of Point Prim Society.

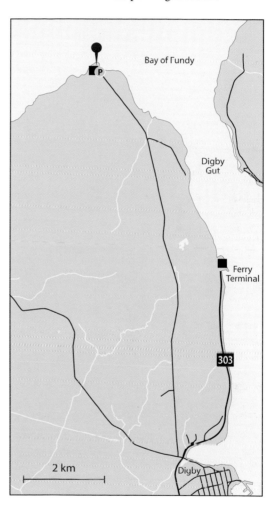

1:50,000 Map

Digby 021A12

Provincial Scenic Route

Evangeline Trail

On the Outcrop

Weathering along vertical column edges emphasizes the polygonal pattern of cracks in the basalt.

Geologists watched and measured a cooling lava lake in Hawaii for several months in the 1960s. Cracks formed on the surface within minutes once the lava stopped flowing. For months as the lava continued to cool, cracks spread downward, creating small seismic tremors and audible snaps and booms, mostly during the cooler hours between 11 p.m. and 3 a.m.

A similar process happened right here, long ago. Look for areas where the basalt is divided by cracks into a tile-like pattern. In some outcrops weathering has revealed how the cracks extend vertically, creating sets of closely packed columns. The columns are typically hexagonal but may have four to eight sides. You may find examples where individual columns cluster to form a larger, approximately round group 1 metre or more across—megacolumns that subdivided as cooling continued.

Silica-rich fluids circulating along the cracks bleached and hardened the margins of some columns. Where weathering has emphasized this effect, the addition of silica has left each column with a rim of harder, more resistant rock encircling a little basin formed by the more easily worn centre.

1000   900   800   700   600   5(

Z1   Z2   Z3   €

FYI

- In a recent study, geologists made detailed measurements of thousands of basalt columns at 50 different European sites. They found that slower cooling of lava leads to a higher proportion of regular, six-sided columns with larger diameters. Faster cooling leads to irregular, small column shapes.

Related Outcrops

From Digby you can follow Highway 217 out Digby Neck to view Nova Scotia's most photographed basalt column, Balancing Rock. From the ferry terminal at Tiverton on Long Island, continue about 4 kilometres to the parking area and trail head (trails.gov.ns.ca/SharedUse/d006.html).

Balancing Rock.

You can also view columnar fractures at Cape d'Or on the Bay of Fundy's northern shore—a spectacular site where Fundy tides appear especially dramatic. Rare deposits of pure copper in the cliffs led to the erroneous name meaning "cape of gold"; the area hosted copper mines long ago.

From Highway 209 in Advocate, follow Back Street briefly, then turn (N45.32870 W64.75850) onto Cape d'Or Road. The road is unpaved; follow it for about 5.5 kilometres to the parking area above the site. From the southwest corner of the parking area, follow the footpath downhill to the open field and explore the outcrops and views as conditions allow.

Cape d'Or.

Exploring Further

The UNESCO World Heritage website, whc.unesco.org/en/list/369, features the Giant's Causeway in Northern Ireland. This is perhaps the world's most famous columnar basalt site, comprising a remarkable 40,000 very regular columns.

A quiet cove near the public wharf for Little Harbour and Cherry Hill hosts a unique but inconspicuous rock formation, part of the Shelburne dyke.

Elusive Giant
Gabbro Dyke of the Central Atlantic Magmatic Province

Picture catching a glimpse of a finback or other large whale as it barely, silently, breaks the water's surface. You don't see much of it, but you can just tell … that's got to be a big one. Of course you want another look at it. So it has been with Nova Scotia's 150-kilometre-long Shelburne Dyke, on average about 120 metres wide.

"Actual outcrops of the dyke have been observed only at two places along this distance … but its location can be easily traced by following the debris, which forms a characteristic red soil, and the weathered … boulders with a rounded pitted surface," wrote Eugene Faribault for a Geological Survey of Canada report in 1919. He had first seen the dyke here at Cherry Hill in 1911 and eight years later had traced it as far as Shelburne.

Since then, this late Triassic intrusion has been traced all the way to Pubnico, south of Yarmouth. It's a big feature and part of a big story, too: The dyke formed when magma poured into a crustal fracture as Pangaea began to break apart.

Getting There

Driving Directions

On Highway 103 (or Highway 3) between Bridgewater and Liverpool, about 14 kilometres northeast of Liverpool, turn (N44.15401 W64.63336) east on Highway 331. Follow Highway 331 for about 14.5 kilometres. Where it curves left, instead fork (N44.15617 W64.48540) right onto Little Harbour Road. You may see signs for Little Harbour wharf. Follow Little Harbour Road for about 250 metres to the parking location.

Where to Park

Parking Location: N44.15785 W64.48333

Park in the gravel lot marked by a harbour authority sign near the cobble beach.

Walking Directions

From the parking area, cross onto the cobble beach and turn right, following the curve of the beach for about 50 metres to a collection of large rounded boulders.

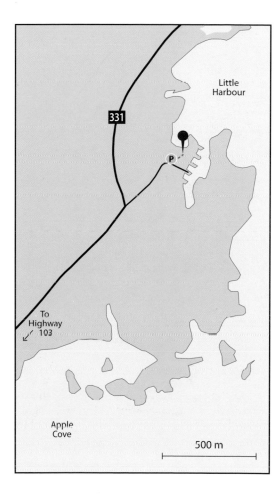

1:50,000 Map

LaHave Islands 021A01

Provincial Scenic Route

Lighthouse Route

On the Outcrop

Rounded outcrops and boulders of the Shelburne Dyke have weathered to a mellow grey-brown. Bright orange areas are patches of lichen.

Outcrop Location: N44.15800 W64.48284

The gabbro outcrops and boulders here are composed mainly of two minerals—white or grey plagioclase and a very dark green, iron- and magnesium-rich mineral, pyroxene. Weathering of the rock has left a reddish or pale brown rind on the outcrops and boulders. Because the pyroxene weathers more quickly than plagioclase, the rind is rough and pitted. This gives the rocks a velvety appearance from a distance.

On some surfaces the weathered exterior has broken away. Where that has happened, you can see the original greenish-grey colour of the fresh gabbro. On close inspection the rock has a speckled appearance due to the light and dark mineral grains.

The rounded boulders here are typical of Shelburne Dyke outcrops. The boulders began as angular blocks, but then weathering affected their corners and edges more quickly than their flat sides. This process, spheroidal weathering, commonly affects igneous rocks having a very even texture.

Weathered and fresh surfaces.

1000   900   800   700   600   50

Z₁   Z₂   Z₃   €

FYI

- Flood basalts (see sites 45 and 46) and related gabbro intrusions affected a large area of Pangaea—about 10 million square kilometres—as the supercontinent began to rift about 200 million years ago. Known as the Central Atlantic Magmatic Province, the affected regions were dispersed by the opening of the Atlantic Ocean and are now found in North and South America, Africa, and Europe.

- The Shelburne Dyke is one of hundreds of similar intrusions in the Central Atlantic Magmatic Province. It may have originally been connected to a similar, 500-kilometre-long dyke in Portugal and Spain.

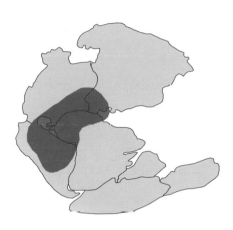

The Central Atlantic Magmatic Province of Pangaea.

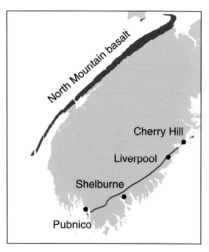

Shelburne Dyke and North Mountain Basalt in western Nova Scotia.

Related Outcrops

The Shelburne Dyke extends across southwestern Nova Scotia from Pubnico past Shelburne, Liverpool, and Cherry Hill, continuing notheastward onto offshore islands and beyond. Subtle chemical differences suggest that it did not serve as a "feeder" for the North Mountain Basalt (see sites 45 and 46), although they are the same age.

Exploring Further

The CAMP site, www.auburn.edu/academic/science_math/res_area/geology/camp/CAMP.html, has a collection of images and information about the igneous province.

00   400   300   200   100   0

€   O   S   D   C   P   Ŧ   J   K   Cz

Nearly horizontal layers of orange-red Jurassic sandstone brighten the shore at Five Islands Provincial Park.

Starting Over Again
Early Jurassic River and Lake Deposits

Jurassic. More than any other geologic time period, it evokes specific, dramatic images, thanks in part to the entertainment industry. Dense foliage munched by thundering herbivores—darting, wily velociraptors … Not far off the mark for late Jurassic times, about 150 million years ago, such scenes earn the period its nickname "age of the dinosaurs."

Hang on, though. Pause to adjust your imagination. The rocks at this site formed quite early in the Jurassic period, about 190 million years ago. Not long before, at the Triassic-Jurassic boundary, nearly half the Earth's known species had become extinct and the region had been smothered in thick sheets of hot lava. The landscape was in recovery mode. Nobody's footsteps thundered, not yet.

The sedimentary features and fossils of these rocks, Nova Scotia's youngest, tell of a changing world. The climate was becoming milder and wetter, but the landscape was still dry much of the time. Plants and animals were repopulating Pangaea, evolving toward the well-known outcome. Something of a cliffhanger, the province's brief but informative Jurassic chapter is aptly nicknamed "dawn of the dinosaurs."

Getting There

Driving Directions

Along Highway 2 about 9 kilometres west of Economy, watch for signs to Five Islands Provincial Park and turn (N45.40771 W64.02163) south onto Bentley Branch Road. Follow the road for about 3 kilometres to the park office. From there, the road dips, then climbs a hill. Near the crest of the hill, fork right onto a gravel road for the day-use picnic area and parking location.

Where to Park

Parking Location: N45.39556 W64.06091

Park in the gravel parking lot for the day-use picnic area.

Walking Directions

From the parking location, walk north down a grassy slope. Where it meets the shore, as conditions allow, cross onto the beach and turn left (south). Walk south following the shoreline for about 375 metres to the outcrop. Continue 350 metres farther for basalt described in By the Way. For a dramatic view farther along the shore, continue 250 metres onto the rocky saddle at The Old Wife. Optionally—as conditions allow—climb across or walk around the saddle to continue southeast along the beach.

Notes

This site lies within the boundaries of Five Islands Provincial Park. The shore is subject to high Fundy tides. Low tide conditions are required both for access and for a safe return. Staff at the park office can advise how best to plan your shoreline walk.

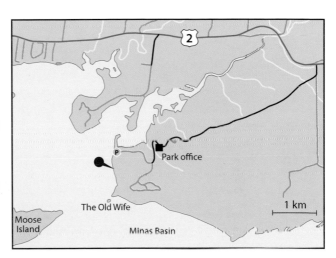

1:50,000 Map

Parrsboro 021H08

Provincial Scenic Route

Glooscap Trail

On the Outcrop

Different weathering rates highlight alternating layers of sandstone and mudstone in the cliff. Near the base, note the thin, lens-shaped area of lake or pond deposits, right of centre (see detail).

Outcrop Location: N45.39309 W64.06159

The thick layers in the cliff were deposited by braided streams—a complex, criss-crossing system of channels that quickly deposited large volumes of sandstone. During floods, widespread layers of mudstone formed as finer sediment settled out slowly.

An eye-catching feature here is a lens-shaped area several metres long near the base of the cliff. The slightly recessed rock is striped with red and grey layers a few centimetres thick. They were deposited in an area of standing water, for example, on the margin of a lake or in a pond that was refilled with flood water from time to time.

Lake or pond deposits.

About 20 metres south of this feature is a sandstone layer containing tablet-shaped fragments of mudstone. Known as rip-up clasts, the fragments formed as fast-flowing, sand-laden water rushed over the surface of an older mud layer, breaking it into pieces.

1000	900	800	700	600	50C
Z₁		Z₂		Z₃	€

FYI

- Nova Scotia's Jurassic rocks contain the richest collection of early dinosaur bones in North America and the oldest known dinosaur bones in Canada (see Exploring Further).

- The Jurassic fossils of Nova Scotia include pollen and spores indicating a variety of vegetation supporting the food chain. Though the Atlantic Ocean had begun to open, land animals could still roam Pangaea: Jurassic fossils found in Nova Scotia have close relatives in the western United States, China, and South Africa.

Related Outcrops

Exposures of Jurassic rock in Nova Scotia are limited to a few small areas along the eastern Bay of Fundy. Between Parrsboro and Five Islands is Wasson Bluff, a site where thousands of Jurassic fossil tracks and bones have been discovered, including those of ancient crocodiles, lizards, and small dinosaurs.

The shore at Wasson Bluff with Five Islands in the distance.

To visit Wasson Bluff, from Parrsboro follow Two Islands Road about 9 kilometres eastward to a gravel parking area (N45.39555 W64.22855) on the south side of the road. The trail head is marked by a sign, "Dawn of the Dinosaurs," including a map of the shore. The site is protected by Nova Scotia's Special Places Protection Act and no collecting is allowed.

Exploring Further

Fundy Geological Museum, 162 Two Islands Road, Parrsboro (N45.39961 W64.32387), fundygeological.novascotia.ca. The museum offers displays and activities (including site tours) related to the volcanic minerals and fossil discoveries of the Parrsboro area, including its remarkable yield of early Jurassic dinosaur fossils.

By the Way

North of The Old Wife, a fault brings sandstone into contact with lava flows of the North Mountain Basalt (see sites 45, 46).

Fault Zone

If you walk beyond the main outcrop at Five Islands and continue south toward The Old Wife, notice the abrupt change in the colour of the rock cliffs. A fault has brought the orange-red Jurassic sandstone into contact with dark grey basalt. As the shore turns southwestward to follow the basalt cliffs, the top of the beach roughly follows the line of the fault. Beside the basalt outcrops, depending on shoreline conditions, you may notice a low pavement of sandstone visible among the beach cobbles underfoot.

Fault breccia.

Some parts of the basalt look as if the rock had been ground up. Irregular chunks are jumbled in a crumbly matrix. This rock type was caused by movements along the fault and is known as fault breccia.

Some original features of the basalt are preserved despite the deformation. Near The Old Wife (for example, N45.38958 W64.06189) are small basalt columns formed as the lava cooled (see site 46). They have been tilted and slightly fractured by fault motion but are still recognizable.

Basalt columns.

By the Way

A basalt lava flow caps this cliff east of The Old Wife. Below it are the light grey and orange-red sedimentary layers onto which it erupted. Greenish-brown fans of debris eroded from the basalt cover parts of the rock face.

Exposed Contact

If conditions allow you to climb up onto the low basalt ridge by The Old Wife at Five Islands (N45.38897 W64.06240) or access the beach farther east (N45.38860 W64.06070), you will come face to face with one of Atlantic Canada's most dramatic geological panoramas.

The colourful lower portion of the cliff face records the accumulation of orange-red Triassic lake and river deposits in the arid Fundy Basin. They are equivalent to similar rocks at Cape Blomidon (site 45). A group of light and dark grey mudstone layers capped the sedimentary sequence, providing the cliff with a bold dash of contrast.

In the upper portion of the cliff are the sombre greys and browns of a basalt flow that poured out onto the sedimentary rock. This flow represents the first of three volcanic episodes, all part of the North Mountain Basalt (see sites 45 and 46).

Summary Timelines

The following two pages provide an overview of your journey through time. The geological timescale from pages 8 and 9 is shown (vertically) for reference, then expanded for each of the book's three sections. Site numbers on the timelines show how the book guides you through the story of Nova Scotia's past (see legend below).

Foundation

The timeline for the Foundation section is complex because the interwoven histories of Avalonia, Ganderia, and Laurentia are complex. Although shown side by side here for comparison, the terranes were separated by tracts of ocean until collisions and fault activity brought them together. Note that some events shown are not described in the book, since relevant sites are difficult to access. Events that affected only part of a terrane are shown as blocks of colour rather than full-width bands.

Meguma

The timeline for the Meguma section is dominated by the long-term accumulation of rock layers in the Goldenville, Halifax, and Rockville Notch groups. Their subsequent metamorphism and intrusion by granites preceded Meguma's oblique collision with Avalonia. The collision caused widespread fault activity across the region.

Pangaea

The timeline for the Pangaea section emphasizes the central role of that supercontinent in Nova Scotia's past. In the late stages of Pangaea's assembly, sediment accumulated in the Maritimes Basin, forming the Horton, Windsor, Mabou, Cumberland, and Pictou groups. After Pangaea's consolidation, uplift of its interior regions left a long silent interval in the province's rock record. The sedimentary rocks and North Mountain Basalt (NMB) of the Fundy Group signal the early stages of Pangaea's break-up.

Timeline Legend

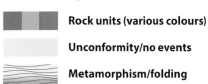

Rock units (various colours)

Unconformity/no events

Metamorphism/folding

Intrusion

Site marker

Foundation

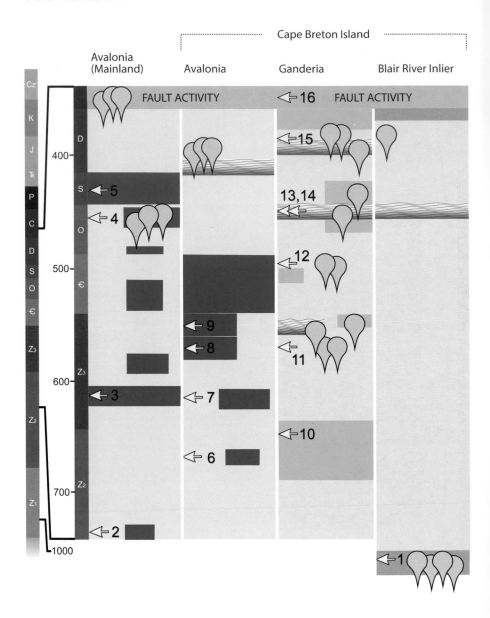

Meguma

Pangaea

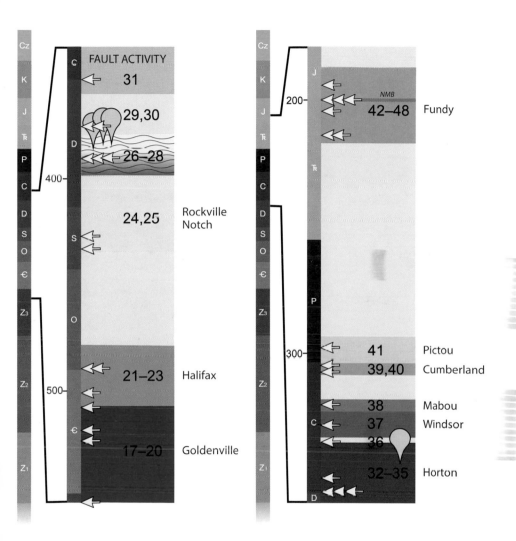

Glossary

acid rock drainage (ARD) A type of pollution formed when water comes into contact with freshly exposed surfaces of sulphur-rich rock (for example, during mining or construction activity), which causes a chemical reaction producing small but damaging amounts of sulphuric acid.

acritarch A microscopic fossil organism with walls of organic material, probably related to plankton but of unproven origin. ACK-rih-tark

algal mound A layered, dome-like accumulation of carbonate rock formed by successive populations of algae living on the bottom of a shallow lake or sea.

alluvial fan A wedge-shaped accumulation of mostly sand and gravel deposited where a stream flows from a mountain valley onto an adjacent lowland.

amphibole A silicate mineral rich in iron and magnesium, typically black or dark green and typically occurring as prismatic crystals in igneous or high-grade metamorphic rocks. It is the dominant mineral in amphibolite, a rock typically formed by metamorphism of mafic intrusions or certain sedimentary rocks. AM-fih-bowl, am-FIB-o-lite

amygdaloidal An igneous rock texture in which a lava matrix contains numerous mineral-filled spheres (amygdales), typically a few millimetres in diameter, which originated as bubbles in the lava. ah-MIG-dah-LOYD ul

andalusite An aluminum silicate mineral, Al_2SiO_5, formed in moderate pressure and temperature conditions. an-de-LOU-zite

andesite A volcanic rock with a composition intermediate between those of rhyolite and basalt. AN-de-zite or AN-de-site

anaerobic Characterized by a lack of oxygen. an-ah-ROE-bic

angular unconformity See unconformity.

anhydrite A calcium sulphate mineral, $CaSO_4$, formed by precipitation from sea water or by dehydration of a related mineral, gypsum. an-HIGH-drite

anorthosite An igneous rock formed primarily from calcium-rich plagioclase feldspar, with small amounts of dark minerals. Most large bodies of anorthosite formed during the Proterozoic eon. ah-NORTH-o-zite

anticline A fold in the shape of an arch, that is, with the hinge or nose of the fold convex upward and the oldest rock layers in the core of the fold. AN-tih-kline

aplite A type of granite having an even, fine-grained texture and very few or no dark minerals. AP-light

arthropod Any member of a class of animals characterized by an exterior skeleton, a segmented body, and jointed legs. ARTH-rah-pod

Avalonia A fragment separated from the continent Gondwana during the Cambrian or Ordovician period and now recognized as a terrane within the Appalachian orogen. Named for the Avalon Peninsula of Newfoundland, where it was first recognized. av-ah-LONE-ee-ah

basalt An extrusive igneous rock formed primarily from plagioclase and pyroxene, identical in composition to gabbro but with mainly fine-grained crystals too small to see with the naked eye. bah-SALT

batholith A complex igneous intrusion at least 100 square kilometres in area. Typically granitic in composition, batholiths may include several distinct but related rock types formed by separate pulses of molten rock. BATH-o-lith

basin A low region of the Earth's surface surrounded by relatively high elevations. They include ocean basins but also basins formed by warping or faulting of continental crust.

bimodal In igneous rocks, characterized by two distinct, contrasting rock types (for example, granite and gabbro) rather than by a single rock type or a range of related types.

bivalve Any member of a class of animals living in salt or fresh water and having a hinged, carbonate shell.

biotite A dark brown or black mica that is rich in iron and magnesium.

black smoker A dark, chimney-like structure that forms on the sea floor where hot brine escapes due to volcanic activity. The brine contains sulphurous and other minerals that build up the chimney over time.

Bouma sequence A five-part sequence of sedimentary layers named for its discoverer, Arnold Bouma, and characterized by diminishing grain size (coarsest layers first, finest last) and a defined series of rock textures. Some parts of the sequence may be missing, but they are never out of order; each sequence records the passage of a single turbidity current. BOOM-ah sequence

brachiopod Any member of a phylum of marine animals having two hinged but unequal shells. BRACK-ee-o-pod

breccia A rock containing a high proportion of angular, broken rock fragments cemented in a finer matrix. BRECH-ee-ah

calcite A calcium carbonate mineral, $CaCO_3$, commonly found filling cracks and other voids in rocks, or making up carbonate rocks such as limestone (sedimentary) or marble (metamorphic).

Canadian Shield A large region of mainly gentle topography in central and northern Canada and the north-central U.S. that has remained geologically stable for at least 1,000 million years.

carbonate A rock consisting of any combination of calcium and magnesium carbonate, including limestone and dolostone (sedimentary) and marble (metamorphic).

chlorite A greenish silicate mineral with a scaly structure, somewhat similar to mica but easily scratched, commonly found in rocks affected by low-grade metamorphism. KLOR-ite

cleavage A rock fabric caused by deformation in which the alignment of flat mineral grains makes the rock weak and easily fractured along any plane parallel to this feature.

conglomerate A sedimentary rock containing a high proportion of rock fragments larger than sand (pebbles, cobbles, or boulders), typically cemented in a finer matrix.

continental shelf The nearshore portion of the continental margin, that is, the shallow marine environment between the shore and the continental slope.

coquina A type of sedimentary rock made entirely, or mostly, of shell fragments. koe-KEEN-ah

cordierite A magnesium-aluminum silicate mineral having a crystal structure similar to beryl. It appears in metamorphosed sedimentary rocks under a wide range of metamorphic conditions. CORD-ee-ah-rite

coticule A metamorphic rock comprising mainly quartz and manganese-rich garnet, typically formed from a manganese-rich muddy siltstone and in some cases used as a whetstone due to its abrasive qualities. COT-ick-yule

crinoid Any member of a class of invertebrate marine animals with flower-like bodies attached to the sea bottom by a long stalk. Crinoids are related to starfish and share their five-fold symmetry. CRY-noid

cross-bedding Inclined deposits of sediment within a larger horizontal bed, formed along the edge of a ripple, bar, dune, or similar feature.

crystal tuff A rock formed by consolidation of volcanic ash containing visible crystals or crystal fragments, usually of quartz and/or feldspar. The crystals formed in a cooling magma chamber beneath the volcano prior to eruption. KRIS-tul TOOF

delta A typically wedge-shaped accumulation of sediment at the mouth of a river, deposited as water flow slows down, causing sediment to settle.

diorite An intrusive igneous rock typically containing two parts plagioclase to one part hornblende and/or pyroxene, and little or no quartz, that is, with a composition intermediate between those of granite and gabbro. DIE-oh-rite

dolostone A carbonate rock similar to limestone, but containing significant amounts of the mineral dolomite, a calcium-magnesium carbonate, $CaMg(CO_3)_2$. DOLE-oh-stone

drag fold A small-scale fold formed in a layer of relatively pliable rock when more rigid layers on either side move in contrasting directions, for example, within a fault zone or along the side of a fold.

ductile Characterized by the ability to change shape without breaking, typically enabled by a combination of factors including high temperature, high pressure, the presence of fluids, and/or favourable mineral content.

dyke A narrow, tabular or sheet-like discordant igneous intrusion formed when molten rock flows into a crack in an existing rock.

epidote A silicate mineral containing calcium, aluminum, and iron, formed during metamorphism of sedimentary rocks or by the alteration of igneous minerals such as pyroxene or amphibole. EPP-ih-dote

euxinia A set of interrelated environmental conditions on the floor of a stagnating lake or sea, characterized by a lack of oxygen and leading to the accumulation of sulphur and carbon in lake- or sea-floor sediments. yooks-IN-ee-ah

exfoliation A complex geologic process, typically affecting igneous rocks of uniform texture, that causes outer layers of nearly uniform thickness to separate from the underlying mass. The process occurs on both large and small scales, leading to rounded shapes whether of entire landforms or individual boulders.

fabric The pattern visible in a rock due to the spatial arrangement and average shape of the mineral grains, rock fragments, or other constituents of the rock.

fault A plane or narrow zone along which a rock mass fractures and displacement occurs. A fault's intersection with the Earth's surface is known as a fault trace, in some cases visible as a landform such as a fault scarp (a steep, linear slope between high and low ground).

feldspar A group of silicate minerals in which silica, aluminum, and varying amounts of sodium, calcium, and potassium combine in a framework-like crystal lattice. Varieties include plagioclase and potassium feldspar.

felsic A rock composition characterized by a predominance of the light-coloured minerals quartz and feldspar.

ferric oxide, ferrous oxide Two forms of iron oxide formed in contrasting conditions. Orange ferric oxide, Fe_2O_3, forms in the presence of plentiful oxygen, while grey ferrous oxide, FeO, forms in oxygen-poor environments.

fluorite A mineral composed of calcium and fluorine, $CaFl_2$, typically purple, blue, or a range of other colours as a result of chemical impurities. Fluorite crystals glow brightly in ultraviolet light.

gabbro An intrusive igneous rock typically containing equal portions of plagioclase and pyroxene. Identical in composition to basalt but with larger crystals easily visible to the naked eye. GAB-roe

Ganderia A fragment separated from the continent Gondwana during the Ordovician period and now recognized as a terrane within the Appalachian orogen. Named for the area around Gander, Newfoundland, where it was first recognized. gan-DEER-ee-ah

garnet A silicate mineral, typically red, formed under conditions of high temperature and pressure, used as a gemstone and as an abrasive.

gneiss A coarse-grained, foliated rock with alternating bands of dark and light minerals. Gneiss can form from either sedimentary or igneous rocks under conditions of high-grade metamorphism. NICE

Gondwana A supercontinent of the geologic past that included areas found in the present-day continents of the southern hemisphere: India, Africa, South America, Australia, and Antarctica. gond-WAH-nah

graded bedding Sedimentary layering in which the size of the particles gradually decreases upward within a single deposit or bed. Grading occurs because the settling rate of a particle in water is proportional to its size (the smaller the particle, the slower the rate).

granite An intrusive igneous rock typically containing equal parts of quartz, plagioclase, and potassium feldspar, with or without small amounts of mica and/or amphibole.

granulite A metamorphic grade characterized by high temperature and pressure, typically occurring 30 kilometres or more beneath the Earth's surface; or, rocks formed under such conditions. GRAN-you-lite

graphite A mineral form of pure carbon in which atoms are arranged in weakly bound sheets. It forms during metamorphism of sediment rich in organic material.

graptolite An extinct class of organism, fossil remains of which are preserved as clusters of black, carbonized impressions, typically in shale. GRAP-to-lite

gypsum A calcium sulphate mineral, $CaSO_4 \cdot 2H_2O$, formed by precipitation from sea water or by hydration of a related mineral, anhydrite. JIP-sum

half-graben A structure formed in rocks under tension, in which one side of a crustal block slips down along a fault, rotating and tilting the block. Typically the movement forms an asymmetrical valley having a steep side along the fault and a gradual slope on the other side. haff GRAH-ben

halite The mineral name for sodium chloride, NaCl. Halite forms in nature when sea water or sodium-rich groundwater evaporates. HAY-light

Iapetus Ocean An ocean of the geologic past that formed as the supercontinent Rodinia split apart, separating Laurentia from Gondwana. ee-APP-ah-tus

inclusion A rock fragment occurring within an igneous rock, having been carried along while the igneous rock was in a molten state and typically originating in rock units through which the magma has travelled.

inlier A defined area of old rock that is surrounded by unrelated younger rock units, typically due to differences in erosion rate, history of uplift, and/or the effects of folding.

intrusion A body of molten rock that has travelled upward through the Earth's crust and invaded or displaced pre-existing rock; or, the process by which this occurs.

island arc A curved chain of islands formed by volcanic activity above a subduction zone.

laminations Millimetre-scale layering in sedimentary rock formed by the deposition of small amounts of material during a succession of minor changes in water conditions or sediment characteristics.

Laurentia A continent of the geologic past (Proterozoic era) that included areas found in present-day North America and parts of Europe. lore-REN-chee-ah

lithosphere The outermost, rigid layer of Earth material, about 100 kilometres thick, consisting of both oceanic and continental crust and the upper mantle, and comprising the Earth's tectonic plates.

lustre A mineral property describing the way light interacts with the surface of a mineral, describing its appearance to the human eye.

magma Molten rock.

magma chamber An underground reservoir of molten rock.

mantle The most voluminous layer of the Earth's interior, lying between the iron-rich outer core and the silica-rich crust.

matrix The fine-grained part of a rock in which larger crystals or rock fragments are embedded.

Meguma The largest of several fragments separated from the continent Gondwana during the Paleozoic era and now recognized as a terrane within the Appalachian orogen. The word is derived from the Mi'kmaq name for their own people. meh-GUE-mah

metamorphism The process by which the minerals in a rock recrystallize in response to changing conditions of temperature and pressure. met-ah-MOR-fizz-em

mica Any one of a group of aluminum-rich silicate minerals having crystals that form stacks or "books" of shiny, thin layers (see biotite, muscovite). MY-cah

microcontinent A fragment of continental crust rifted from a larger, pre-existing continental mass.

migmatite A rock formed by partial melting during intense metamorphism and typically consisting of a dark matrix containing lighter-coloured bands and pods that were once molten.

mudstone A sedimentary rock that originated as very fine silt and clay and that lacks the easily parted layering of shale.

muscovite A type of mica lacking iron or magnesium, recognized by its transparent or silvery colour. MUSK-ah-vite

mylonite A fine-grained rock type characterized by strong linear and/or planar fabric formed during metamorphism and intense deformation in a fault zone. MY-lah-nite

nuée ardente Literally, "glowing cloud"; a type of volcanic eruption characterized by fast-moving, turbulent masses of hot volcanic gas, ash, and debris that flow down the mountainside. NOO-ay ar-DAHNT

oceanic ridge The long, narrow, mountainous area of an ocean basin along which new ocean crust is formed by a continuous process of rifting and volcanic activity.

orogeny The process of mountain building. An orogen is the result of this process as preserved in the rock record, regardless of whether the region is still mountainous or has been worn down by erosion. oh-RODGE-enn-ee, ORE-oh-jen

oxidation A chemical process resulting in the addition of oxygen to certain molecules, for example, during the weathering of minerals or the decomposition of organic matter.

paleomagnetic Pertaining to the alignment of magnetic minerals in a rock at the time the rock formed. Because magnetic minerals tend to align with the Earth's magnetic field, paleomagnetic data provide information about where on Earth the rock was located when it formed. PAY-lee-o-mag-NET-ick

paleosol A layer of ancient soil preserved in its original context (above the rock on which it formed) as part of the rock record. PAIL-ee-oh-SOLL

Pangaea A supercontinent of the geologic past (approximately 300 to 175 million years ago). pan-GEE-ah

passive margin A continental margin situated adjacent to old, stable oceanic crust not subject to significant tectonic activity; not a tectonic plate boundary.

pegmatite An igneous rock type characterized by large (centimetre-scale or greater) mineral grains, typically occurring in veins or pods within an igneous intrusion and in some cases hosting rare minerals containing boron, lithium, or fluorine. PEG-mah-tight

peri-Gondwanan, peri-Laurentian Of or relating to terranes originating on or near the continents of Gondwana or Laurentia, respectively.

phyllite A rock type similar to slate, but formed in more intense conditions of heat and pressure and thus having larger grains of mica, which give the rock a silky appearance. FILL-ite

pillow lava A form of basalt exhibiting bulbous shapes formed as lava erupts under water. Some pillows shatter during this process, forming pillow breccia.

plagioclase A type of feldspar containing calcium and/or sodium rather than potassium. PLAJ-ee-o-klaze

playa A shallow lake from which water does not flow because the rate of evaporation matches or exceeds the rate of input from rain or river drainage; or the flat expanse of salty soil left when the lake periodically evaporates. PLY-uh

poikiloblast A crystal formed during metamorphism that engulfed and preserved smaller pre-existing crystals as it grew. POIK-ih-low-blast

potassium feldspar A type of feldspar containing potassium rather than calcium and/or sodium. Potassium feldspar is typically pinkish and lends granite its characteristic colour.

pseudotachylite A rock formed by instantaneous melting in response to shock, for example during movement on a fault or a meteorite impact. Due to its glassy texture, it is typically darker than the surrounding unaltered rock. soo-doe-TACK-ih-lite

pyrite A metallic mineral, iron sulphide, also known as "fool's gold" due to its shiny, yellow crystals. PIE-rite

pyroclastic Characterizing volcanic eruptions dominated by the violent ejection of hot ash, droplets of magma, rock fragments, and other material. pie-roe-CLASS-tick

pyroxene A silicate mineral rich in iron and magnesium, similar to amphibole but with a simpler crystal structure that forms at higher temperatures. PEER-ox-een

Rheic Ocean An ocean of the Paleozoic era that formed between Gondwana and several microcontinents, including Ganderia, Avalonia, and Meguma. REE-ick or RAY-ick OH-shun

rift A regional-scale break in the Earth's crust, caused by tectonic forces causing tension within oceanic or continental crust, typically resulting in deep, steep-sided valleys and, in some cases, volcanic activity.

rhyolite A volcanic rock equivalent in composition to granite. RYE-o-lite

Rodinia A supercontinent of the geologic past (approximately 1,000 to 750 million years ago). roe-DIN-ee-ah

schist A strongly foliated rock rich in plate-like minerals such as mica. Schist can form by metamorphism of either sedimentary or igneous rocks. SHIST

sea stack A coastal landform characterized by a tall, isolated, steep-sided rock adjacent to the shoreline, formed by erosion acting on vertically oriented weaknesses in the rock.

shale A sedimentary rock type formed from muddy sediment and having a flaky texture due to the alignment of clay minerals.

shear zone Similar to a fault in that both are caused by relative movement of adjacent blocks of crust, but in a shear zone, the movement is "smeared out" across a wide band instead of being focused along a single plane.

siderite An iron carbonate mineral, $FeCO_3$, typically found in hydrothermal veins or sedimentary rocks and valued as an ore due to its lack of sulphur. SID-uh-rite

siltstone A sedimentary rock formed from silt, that is, sediment with a grain size between that of sand and mud.

staurolite An iron-aluminum silicate mineral characterized by dark red or brown colour, extreme hardness, and intergrowth of crystals to form various cross-like shapes; one of several minerals used to estimate metamorphic temperature and pressure conditions. STORE-uh-lite

stratovolcano A type of volcano, typically tall and steep-sided, formed by the accumulation of layers of ash, other volcanic debris, and lava.

striation In glaciated landscapes, one of a set of usually parallel gouges or scratches on the surface of a rock, caused by glacial abrasion as rock-laden ice scrapes across a rocky landscape. stry-AY-shun

subduction A process by which one tectonic plate moves beneath another along their common boundary and sinks into the Earth's mantle. The setting where this occurs is known as a subduction zone. sub-DUCK-shun

supercontinent A landmass that includes all or many regions of the Earth's continental crust, assembled through a series of continental collisions.

syenite An intrusive igneous rock, typically brownish-red or dark pink in colour, dominated by potassium feldspar and including smaller amounts of plagioclase but little or no quartz. SIGH-uh-nite

tempestite A sedimentary rock characterized by signs of water turbulence and deposited in shallow coastal environments during stormy conditions. tem-PEST-ite

terrane A fault-bounded fragment of continental crust broken from one tectonic plate and later joined to another during continental collision, recognized as such by its distinctive rock units and separate sedimentary, igneous, and/or metamorphic history as compared to adjacent parts of an orogen.

texture A characteristic of rocks that describes the size, shape, and arrangement of its constituent parts.

tonalite An igneous rock containing primarily quartz and plagioclase, with lesser amounts of dark minerals such as biotite and amphibole. TONE-ah-lite

tourmaline A semi-precious boron silicate mineral found in a wide variety of colours due to impurities within the crystal structure. A common black variety is known as schorl. TOUR-mah-leen

trace fossil Preserved evidence of animal activity, for example, in the form of tracks, burrows, feeding marks, or resting places.

transcurrent fault A fault along which opposing crustal blocks move sideways past one another.

tuff A rock formed by consolidation of volcanic ash deposited either on land or into a body of water from which it later settled. TOOF

turbidite A sedimentary rock deposited in deep ocean water by turbulent, avalanche-like currents of liquefied sediment (known as turbidity currents) flowing down the continental slope. TUR-bah-dite

unconformity An erosion surface preserved in a sequence of rock layers, representing a period of time during which no sediment was deposited or, if deposited, was subsequently removed by erosion. In an angular unconformity, sedimentary layers below the unconformity are tilted with respect to the younger layers, signifying an intervening cycle of deformation, uplift, and subsidence. un-con-FOR-mit-tee

vent A crack or pipe-like opening through which lava, steam, or other volcanic emissions reach the Earth's surface.

volcanic arc A linear or slightly curved pattern of volcanic activity above a subduction zone. A volcanic arc located entirely within oceanic crust is called an island arc.

volcanic ash A collective term used to refer to fine-grained fragments (less than 2 millimetres wide) of volcanic glass, crystals, and rock ejected during an explosive eruption.

volcanic block, volcanic bomb Large fragments ejected during a volcanic eruption; blocks are angular fragments of pre-existing rock, and bombs form when liquid magma cools and solidifies as it falls.

volcanogenic massive sulphide (VMS) deposit A type of mineral deposit dominated by sulphides of copper and zinc, formed on the ocean floor by the circulation of hot, mineral-rich fluids along mid-ocean ridges and other ocean-floor rifts.

zeolite A group of aluminum silicate minerals having a porous crystal structure and that typically form as groundwater circulates through volcanic rock or deposits of volcanic ash. The ability of zeolite crystals to absorb and filter other molecules makes them useful for a variety of industrial applications. ZEE-oh-lite

zircon A zirconium silicate mineral, $ZrSiO_4$, commonly formed as tiny crystals in felsic igneous rocks and useful for measurement of geologic time due to their uranium content.

Index of Place Names

Image Credits

All photographs, maps, and diagrams are by Martha Hickman Hild and Sandra M. Barr, except as noted below:

Page 18, mountain belts map adapted from van Staal & Barr, 2012, *Geological Association of Canada Special Paper 49*, fig. 1 (p. 44).

Page 49, paleomagnetic diagram adapted from Hodych & Buchan, 1998, *Geophysical Journal International*, fig. 5 (p. 160) with additional data from Thompson, Grunow, & Ramezani, 2010, *Geological Society of America Bulletin*, vol. 119, p. 681.

Page 120, 121, photos (3) of Mn-rich layers and coticules by Chris White. Published with permission.

Page 178, photo of folded sedimentary rock by Jonathan Shute. Published with permission.

Page 215, photo of Prince Edward Island by Rob Fensome. Published with permission.

Page 221, photos (2) of Rainy Cove folds by Alan Macdonald. Published with permission.

Page 225, Triassic basins diagram adapted from Hamblin, 2004, *Geological Survey of Canada Open File 4678*, fig. 2 (p. 27).

Page 229, landforms diagram adapted from LeLeu & Hartley, 2010, *Journal of the Geological Society of London*, vol. 167, fig. 7B (p. 448), with assistance and permission from Dr. Sophie LeLeu.

Page 237, photo of Balancing Rock by Rob Fensome. Published with permission.

Road maps were produced using datasets obtained online from GeoGratis (geogratis. cgdi.gc.ca) under the Open Government Licence—Canada (open.canada.ca/en/open-government-licence-canada). Although every effort has been made to provide helpful directions based on the datasets, under the terms of the licence, no representation or warranty of any kind is made with respect to the accuracy or completeness of the information. Please consult the licence terms for additional details.

About the Authors

Dr. Martha Hickman Hild received her PhD in Earth Sciences in 1976 from the University of Leeds, UK. Early in her career, she co-directed a research laboratory and lectured in geology. Subsequently she worked as an editor in both technical and educational publishing and as an award-winning news researcher. She is now a freelance writer and editor. The innovative format and writing style she developed for her first book, *Geology of Newfoundland*, led to its selection as "Best Guidebook 2014" by the Geosciences Information Society.

Dr. Sandra M. Barr received her PhD in Geology in 1973 from the University of British Columbia. Since 1976, she has been teaching geology at Acadia University in Wolfville, NS, where she is a Professor in the Department of Earth and Environmental Science. Widely recognized for her field-based studies of Appalachian geology, she has authored and co-authored 200 scientific publications and produced numerous reports and maps, many focused on Nova Scotia's rocks. She is book editor for the Geological Association of Canada and co-editor of the journal *Atlantic Geology*.